CONTENTS

	Page
Hearing held on March 22, 2002	1
Statement of:	
Barnes, Johnny, executive director, American Civil Liberties Union, National Capital Area; Ronald Goldstock, American Bar Association; and John Woodward, Jr., RAND	57
Patterson, Kathleen, chairman, Committee on the Judiciary, Council of the District of Columbia; Margret Nedelkoff Kellems, deputy mayor, Public Safety and Justice, District of Columbia government; Charles Ramsey, chief, Metropolitan Police Department, District of Columbia; and John Parsons, associate regional director, National Capital Region, National Service	13
Letters, statements, etc., submitted for the record by:	
Barnes, Johnny, executive director, American Civil Liberties Union, National Capital Area, prepared statement of	60
Goldstock, Ronald, American Bar Association, prepared statement of	69
Kellems, Margret Nedelkoff, deputy mayor, Public Safety and Justice, District of Columbia government, prepared statement of	40
Morella, Hon. Constance A., a Representative in Congress from the State of Maryland, prepared statement of	4
Norton, Hon. Eleanor Holmes, a Delegate in Congress from the District of Columbia, prepared statement of	10
Parsons, John, associate regional director, National Capital Region, National Service, prepared statement of	33
Patterson, Kathleen, chairman, Committee on the Judiciary, Council of the District of Columbia, prepared statement of	16
Ramsey, Charles, chief, Metropolitan Police Department, District of Columbia, prepared statement of	24
Woodward, John, Jr., RAND, prepared statement of	78

PRIVACY VS. SECURITY: ELECTRONIC SURVEILLANCE IN THE NATION'S CAPITAL

FRIDAY, MARCH 22, 2002

HOUSE OF REPRESENTATIVES,
SUBCOMMITTEE ON THE DISTRICT OF COLUMBIA,
COMMITTEE ON GOVERNMENT REFORM,
Washington, DC.

The subcommittee met, pursuant to notice, at 9:05 a.m., in room 2154, Rayburn House Office Building, Hon. Constance A. Morella (chairwoman of the subcommittee) presiding.

Present: Representatives Morella and Norton.

Staff present: Russell Smith, staff director; Heea Vazirani-Fales, counsel; Matthew Batt, legislative assistant/clerk; Robert White, communications director; Shalley Kim, staff assistant; Howie Denis, professional staff member (Davis); Jon Bouker, minority counsel; and Earley Green, minority assistant clerk.

Mrs. MORELLA. Good morning. I am going to call the Subcommittee of the District of Columbia hearing to order. Our issue today is privacy versus security, electronic surveillance in the Nation's Capital. I want to welcome everybody who is here.

I am going to ask that my opening comment, in its entirety, be included in the record because I am going to give an abbreviated version. Thank you. Without objection.

We live in the video age. Police forces, including the Metropolitan Police Department, are increasingly employing video surveillance, both to deter crime and to catch criminals. The Metropolitan Police Department is in the process of establishing the most extensive surveillance network in the United States; a system that could ultimately include more than 1,000 cameras, all linked to a central command station accessible to not only the District police but the FBI, the Capitol Police, the Secret Service, and other law enforcement agencies.

The existence of such a network raises many questions. Among them, does the prevalence of cameras inhibit our privacy rights? Are those cameras effective in deterring or solving crimes? And, perhaps most urgently, who gave permission for the implementation of this system, and where are the policies governing its use?

I believe there has been an unfortunate lack of public debate on these issues. Even supporters of electronic surveillance concede that police departments should only use these cameras if there is a widespread public desire for such technology.

There is clearly no consensus the District of Columbia for or against these cameras, because the public only learned about their existence after they had been put in place. Citizens must have confidence that electronic surveillance is not going to infringe on their rights, including what Justice Louis Brandeis described as our most precious rights, the right to be left alone.

We saw the dangers of moving too quickly with this technology when the District had its problems with the faulty red light cameras, and the due process issues now being raised regarding the speeding cameras.

I understand the Metropolitan Police Department now has 13 closed circuit cameras of its own linked to its joint operations command center, and it is working on linking the center to several hundred existing cameras in public schools and subway stations.

There are also plans to connect the center with hundreds of regional natural traffic cameras. One of the biggest concerns that I have is that once this system is in place, it will be too tempting for the police not to use it to its full force. It is the old camel's nose story, or maybe we should say the camera's nose. Once the camel gets his nose under the tent, pretty soon the rest of the camel will be under the tent. And once the police have cameras that can see anywhere in the city, pretty soon the police will be using those cameras to look anywhere in the city.

In London, a camera system initiated to combat IRA terrorism has sprouted into a network with an estimated 2½ million cameras. The average Londoner is caught on film about 300 times a day, and no terrorists have been caught by the cameras' use.

Does the Nation's Capital want to build such a system? I have heard Chief Ramsey say no, that this is a system designed to be event-specific, to be activated only during threats of terror or large public events. But Mayor Williams had said publically that the city should follow the lead of cities such as London which use the cameras to enhance day-to-day policing. The chief has said the Department will consider installing cameras in neighborhoods that have problems with drug markets and the like.

Obviously, some guidance, maybe legislation from the Mayor and the Council of the District of Columbia is needed to establish how extensive this camera network should be and what safeguards are necessary to protect privacy rights.

The policymakers must give clear direction to the police. Congress, too, may have to step in and ensure that this technology does take—does not take away our right to be left alone, as it has in the past when privacy concerns have become an issue. This is especially true given the testimony from the U.S. Park Service, which is planning to place cameras at the monuments on our National Mall.

We have two panels today, one that will primarily focus on the District's own surveillance system, and one that will be able to broaden the discussion a little further into the Constitutional and legislative questions. We did invite several others to testify, including the Capitol Police and Justice Department, but they declined, saying they had no role to play in the discussion of the District's surveillance network. And the British Government doesn't let government officials testify before Congress, so we were unable to get someone who could speak with firsthand knowledge about London's experience.

In concluding my remarks, I wanted to say that our Nation's Capital stands as an ultimate symbol to American freedom. Since taking the chairmanship of this subcommittee 15 months ago, I have worked with Congresswoman Norton on many issues, but perhaps no single one as frequently as trying to keep this city safe, open and accessible to the residents, businesses, and 19 million tourists who come here each year.

I have said before that we cannot turn the District into Ft. Washington. It matters not whether that fortress is built with an impenetrable ring of concrete barriers or with an unregulated network of digital cameras.

It is now my pleasure to recognize the ranking member, Congresswoman Eleanor Holmes-Norton, for her opening statement.

[The prepared statement of Hon. Constance A. Morella follows:]

CONSTANCE A. MORELLA
8TH DISTRICT, MARYLAND

COMMITTEE ON GOVERNMENT REFORM
CHAIR,
SUBCOMMITTEE ON THE DISTRICT OF COLUMBIA
SUBCOMMITTEE ON CIVIL SERVICE

COMMITTEE ON SCIENCE
SUBCOMMITTEE ON ENVIRONMENT, TECHNOLOGY, AND STANDARDS

Congress of the United States
House of Representatives

2228 RAYBURN HOUSE OFFICE BUILDING
WASHINGTON, DC 20515
(202) 225-5341
FAX: (202) 225-1389

www.house.gov/morella

51 MONROE STREET
SUITE 507
ROCKVILLE, MD 20850
(301) 424-3501
FAX: (301) 424-5992

CHAIRWOMAN CONSTANCE A. MORELLA
HOUSE DISTRICT OF COLUMBIA SUBCOMMITTEE
OPENING STATEMENT
"PRIVACY VS. SECURITY: ELECTRONIC SURVEILLANCE IN THE NATION'S CAPITAL"
MARCH 22, 2002

We live in a video age. Information from all over the world is beamed in real time across any number of television stations; the first instinct of many eye witnesses to breaking news is to grab their camcorders and become amateur reporters; the places where we shop, eat and bank all use security cameras.

Naturally, police forces, including the Metropolitan Police Department here in the Nation's Capital, are increasingly employing video surveillance, both to deter crime and to catch criminals. In our Nation's Capital, the MPD is in the process of establishing the most extensive surveillance network in the United States – a system that could ultimately include more than 1,000 cameras[1], all linked to a central command station accessible to not only the District police, but the FBI, the Capitol Police, the Secret Service, and other law enforcement agencies.

The existence of such a network raises many questions. Does the prevalence of cameras inhibit our privacy rights? Are these cameras effective in deterring or solving crimes? When and how will these cameras be used? Who controls them? Who has access to the recorded images? How long are these images retained, and for what purposes? And, perhaps most urgently, who gave permission for the implementation of this system, with cameras primarily pointing at federal government buildings, and where are the policies governing its use?

I called this hearing because I believe there has been an unfortunate lack of public debate on these issues. Even supporters of electronic surveillance – including the Security Industry Association and the International Association of Chiefs of Police – concede that police departments should only use these cameras if there is a widespread public desire for such technology. There is clearly no consensus in the District of Columbia for or against these cameras, because the public only learned about their existence after they had been put in place.

---

[1] The Metropolitan Police Department has 13 cameras. D.C. Public Schools and the Metro subway system have more than 200 cameras each. The D.C. Department of Transportation wants to establish a network of 700 traffic cameras. All would have the ability to be linked to the central command center.

Citizens must have confidence that electronic surveillance is not going to infringe on their rights – including what Justice Louis Brandeis described as our most precious right, "the right to be left alone." We saw the dangers of moving too quickly with this technology, and without proper oversight, when the District had its problems with faulty red-light cameras, and the due process issues now being raised regarding the speeding cameras. Several constituents of mine have received speeding tickets from these cameras for cars they had previously sold. One of them, in fact, is an elderly woman who lives in a nursing home and no longer drives. Imagine her surprise at getting ticketed for speeding.

I understand the Metropolitan Police Department now has 13 closed-circuit cameras of its own linked to its Joint Operations Command Center, and it is working on linking the center to several hundred existing cameras in public schools and subway stations. There are also plans to connect the center with hundreds of regional traffic cameras.

One of the biggest concerns I have is that, once this system is in place, it will be too tempting for the police not to use it to its full force. It's the old camel's nose story – or maybe we should say the *camera's* nose. Once the camel gets his nose under the tent, pretty soon the rest of the camel will be under the tent. Once the police have cameras that can see anywhere in the city, pretty soon the police will be *using* those cameras to look anywhere in the city.

We can look at London for guidance. A camera system initiated to combat IRA terrorism has sprouted into a network with an estimated 2-and-a-half million cameras. The average Londoner is caught on film about 300 times per day. And, although the British police believe the cameras have helped reduce crime, no terrorists have been caught by their use.

Does the Nation's Capital want to build such a system? I have heard Chief Ramsey say no, that this is a system designed to be "event-specific," to be activated only during threats of terror or large public events. But Mayor Williams has said publicly that the city should follow the lead of cities such as London, which use the cameras to enhance day-to-day policing. The Chief has also said the department will consider installing cameras in neighborhoods that have problems with drug markets and the like.

Obviously, some guidance – and legislation – from the Mayor and the Council of the District of Columbia is needed to establish how extensive this camera network should be and what safeguards are necessary to protect privacy rights. The policy makers must give clear direction to the police about how video surveillance can be used. Written agreements must be forged with others who might use the Joint Command Center, whether they are federal law enforcement agencies or other local police forces.

As I learn more about this issue, it is becoming more evident that Congress may have to step in and ensure that this technology does take away our "right to be left alone." This is especially true given the testimony we will hear today from the U.S. Parks Service, which is planning to place cameras at the monuments on our National Mall.

Because the Constitution does not explicitly grant a right to privacy, it has been often left to Congress to draw the boundaries. In 1974, for example, Congress passed the Privacy Act, which set rules for federal agencies on releasing personal information. Two years later, Congress established minimum privacy requirements for banks and financial institutions, leaving the states free to enact tougher standards if they so desired. That could well be something for us to consider concerning video surveillance.

We have two panels today, one that will primarily focus on the District's own surveillance system and one that will be able to broaden the discussion a little further into the Constitutional and legislative questions. We did invite several others to testify, including the Capitol Police and Justice Department, but they declined, saying they had no role to play in a discussion of the District's surveillance network. And the British government does not let government officials testify before Congress, so we were unable to get someone who can speak with first-hand knowledge about London's experience.

In concluding my remarks, I'd like to say that our Nation's Capital stands as the ultimate symbol to American freedom. Since taking the chairmanship of this Subcommittee 15 months ago, I have worked with Congresswoman Norton on many issues, but perhaps no single one as frequently as trying to keep this city safe, open and accessible to the residents, businesses and 19 million tourists who come here each year. I've said before that we cannot turn the District into "Fort Washington" – and it matters not whether that fortress is built with an impenetrable ring of concrete barriers or with an unregulated network of digital cameras.

#####

Ms. NORTON. Thank you, Ms. Morella. I appreciate the leadership of our chair, Connie Morella, in focusing our subcommittee on important issues of privacy raised by security threats and technology that have been become a special concern since the September 11th attack.

The surveillance cameras at issue were initiated before September 11th, however. Initially they were not used because of terrorist threats to security, but for law enforcement purposes at national events where lawlessness from demonstrators and others sometimes occurs.

Thus, it seems clear that as technology becomes available, government, like the rest of us, gravitate toward its use. However, government is not like the rest of us. Government is not like citizens with a new toy. There are deeply felt cultural norms and critical constitutional limits on how public officials may carry out their work, even responsibilities as important as public safety. For Americans, where cameras that view residents should be placed is not merely a security issue. Personal privacy and the right to be left alone, especially from interference by government, is an identifying characteristic of what it means to be an American.

Nevertheless, much of what we have focus on today is uncharted territory. Our subcommittee is one of the first in Congress to investigate surveillance cameras that are used for security purposes, perhaps the very first.

Such cameras in the District present a particularly difficult case for arriving at the appropriate balance in a society like ours. As the Nation's Capital, the District is a presumed terrorist target. And as the use of the surveillance cameras here, even before September 11th, shows, the city also is the formal site for many national and international events that have the potential to generate harm to residents and visitors and damage to property.

The cameras are now connected to the District's joint operations command center. And I want to especially commend council member Kathy Patterson, chair of the City Council's Committee on the Judiciary, for promptly calling witnesses and for planning further hearings concerning the surveillance cameras. The city's first hearing has uncovered important information and already is leading to remedial action.

This subcommittee has larger Federal concerns that affect not only the District, but the Nation. Even the District's use of surveillance cameras was motivated not by District matters, but by security needs at national and international events. And already the District system is trending toward other Federal uses that may rapidly become significantly greater than local involvement.

Initially at least five of 13 locations are to Federal sites. Already three more Federal agencies may be seeking connections. There is no way to avoid the conclusion that in the new era of global terrorism, the District's cameras surveillance system is inevitably already part and parcel of the Nation's homeland defense.

As such, the surveillance is likely to be imitated in other locations, especially in the many jurisdictions where there is a Federal presence. The need to assure greater security in the Nation's Capital and elsewhere, especially following September 11th, is beyond debate.

Today, however, we will want to know more about the uses to which the system is being and could be put, how and when the system is used, its limits, its known benefits and dangers, and what the surveillance has accomplished so far.

As perhaps the first such surveillance system in the country, there is a heavy burden on the users to set the appropriate example and to do it right. However, public officials here are caught in a dilemma not of their own making. Like other officials throughout the country, they are being asked to respond to the unknown with no precedence to guide them.

I believe that it is both unfair and dangerous for the national government to ask local and State officials to figure out the plethora of complicated issues involving security, privacy and openness no society in the history of the world has had to face without any Federal guidance.

However, the Federal Government itself faces the same complicated challenge, to protect the people while maintaining their constitutionally guaranteed rights. The example that the Federal Government is offering leaves much to be desired. Federal officials are quickly throwing up new approaches and systems from shutdowns, barricades and public exclusion to camera surveillance, wholesale round-ups and electronic surveillance, without regard to the Fourth Amendment or other constitutional protections, and without notice and public hearings or even any public explanation.

Civil libertarians are right to question much about this response, and public officials are right to revitalize their responsibility to protect the public with new and increased seriousness.

However, both the general public and public officials need more explicit guidance from the Federal Government. Yet Homeland Security Director, Tom Ridge, has all he can do simply to catch up to an increasing set of new demands created by September 11th, most of them just beginning to take on the basics, such as calibrated alerts, and the security safeguards that are necessary to fully reopen National Airport.

Because both the local and Federal Governments face the same dilemma, it is particularly unjustifiable that nearly all of the Federal officials invited to testify today have declined. They are the Office of Homeland Security, the Department of Justice, the Federal Bureau of Investigation, the Capitol Police, and the U.S. Secret Service, particularly since most of the new security needs or requirements are Federal in origin.

The least that Congress is entitled to is the kind of testimony that can be presented without injury to national security. When even that testimony is withheld, complaints from administration agencies concerning any legislation that results will be unavailing. I believe that a more cooperative and forthright approach that faces these dilemmas head-on together trying to find solutions is what is needed to help sort out the conundrum of at once opening the society and closing down terrorism.

I will shortly be introducing the Open Society With Security Act, along with a Senate sponsor. The bill would authorize a Presidential commission to bring together the best minds in the society to investigate how our country can meet the high standards necessary to effectively fight the dangerous menace of international terrorism while accommodating and affirming the central American values of privacy, openness and public access.

Like the Kerner Commission, the Open Society With Security Commission would help us to chart a safe course through deep waters, without surrendering the very values that lead us to insist upon defending our country and our way of life.

We can do better than blunt and often untested and ineffective instruments that crush our liberty. I spent my early years at the bar as assistant legal director for the national ACLU, and today I have responsibilities for national security as a Member of Congress. My experience in both roles has reinforced my confidence that American ingenuity is ready for the new challenge of winning the struggle against dangerous and dogmatic terrorism, while maintaining and enriching the free and open democratic society that virtually defines our country.

I welcome today's witnesses and appreciate their testimony. I am certain that when we have heard them, all of us will be in a better position to find the appropriate solutions. Thank you, Madam Chair.

[The prepared statement of Hon. Eleanor Holmes Norton follows:]

ELEANOR HOLMES NORTON
DISTRICT OF COLUMBIA

COMMITTEE ON
TRANSPORTATION AND
INFRASTRUCTURE

SUBCOMMITTEES

AVIATION

ECONOMIC DEVELOPMENT, PUBLIC
BUILDINGS, AND EMERGENCY
MANAGEMENT

COMMITTEE ON
GOVERNMENT REFORM

SUBCOMMITTEES

RANKING MINORITY MEMBER,
DISTRICT OF COLUMBIA

CIVIL SERVICE AND
AGENCY ORGANIZATION

Congress of the United States
House of Representatives
Washington, D.C. 20515

Statement of Congresswoman Eleanor Holmes Norton
District of Columbia Subcommittee Hearing on
"Privacy v. Security: Electronic Surveillance in the Nation's Capital"

March 22, 2002

I appreciate the leadership of our chair, Connie Morella, in focusing our subcommittee on important issues of privacy raised by security threats and technology that have become a special concern since the September 11 attacks. The surveillance cameras at issue were initiated before September 11, however. Initially, they were not used because of terrorist threats to our security, but for law enforcement purposes at national events where lawlessness from demonstrators and others sometimes occurs. Thus, it seems clear that as technology becomes available, government, like the rest of us, gravitates toward its use. However, government is not like citizens with a new toy. There are deeply felt cultural norms and critical constitutional limits on how public officials may carry out their work, even responsibilities as important as public safety. For Americans, where cameras that view residents should be placed is not merely a security issue. Personal privacy and the right to be left alone, especially from interference by government, is an identifying characteristic of what it means to be an American. Nevertheless, much of what we focus on today is unchartered territory. Our subcommittee is one of the first in Congress to investigate surveillance cameras that are used for security purposes.

Such cameras in the District present a particularly difficult case for arriving at the appropriate balance in a society like ours. As the nation's capital, the District is a presumed terrorist target, and, as the use of the surveillance cameras here even before September 11 shows, the city also is the foremost site for many national and international events that have the potential to generate harm to residents and visitors and damage to property. The cameras are now connected to the District's Joint Operations Command Center, and I want to especially commend Council member Kathy Patterson, chair of the City Council's Committee on the Judiciary, for promptly calling witnesses and for planning further hearings concerning the surveillance cameras. The city's first hearing has uncovered important information and already is leading to remedial action.

This subcommittee has larger federal concerns that affect not only the District, but national concerns. Even the District's use of camera surveillance was motivated by security needs at national and international events, and, already, the District's system is trending toward other federal uses that may rapidly become significantly greater than local involvement. Initially, at least five of thirteen locations are related to federal sites. Already, three more federal agencies

815 15TH STREET, N.W., SUITE 100
WASHINGTON, D.C. 20005-2201
(202) 783-5065
(202) 783-5211 (FAX)

2136 RAYBURN HOUSE OFFICE BUILDING
WASHINGTON, D.C. 20515-5101
(202) 225-8050
(202) 225-3002 (FAX)
(202) 225-7829 (TDD)
www.house.gov/norton

2041 MARTIN L. KING AVENUE, S.E.
SUITE 300
WASHINGTON, D.C. 20020-5734
(202) 678-8900
(202) 678-8844 (FAX)

may be seeking connections. There is no way to avoid the conclusion that in the new era of global terrorism, the District's camera surveillance system is inevitably already part and parcel of the nation's Homeland defense. As such, the surveillance is likely to be imitated in other locations, especially in the many jurisdictions where there is a federal presence.

The need to assure greater security in the nation's capital and elsewhere, especially following September 11, is beyond debate. Today, however, we will want to know more about the uses to which the system is being and could be put, how and when the system is used, its limits, its known benefits and dangers, and what the surveillance has accomplished so far. As perhaps the first such surveillance system in the country, there is a heavy burden on users to set the appropriate example and to do it right.

However, public officials here are caught in a dilemma not of their own making. Like other officials throughout the country, they are being asked to respond to the unknown with no precedents to guide them. I believe that it is both unfair and dangerous for the national government to ask local and state officials to figure out the plethora of complicated issues involving security, privacy, and openness no society in the history of the world has had to face, without any federal guidance.

However, the federal government itself faces the same complicated challenge: to protect the people while maintaining their constitutionally guaranteed rights. The example that the federal government is offering leaves much to be desired. Federal officials are quickly throwing up new approaches and systems – from shutdowns, barricades, and public exclusion, to camera surveillance, wholesale round-ups and electronic surveillance, without regard to the Fourth Amendment or other constitutional protections, and without notice and public hearings, or even any public explanation.

Civil libertarians are right to question much about this response, and public officials are right to revitalize their responsibility to protect the public with new and increased seriousness. However, both the general public and public officials need more explicit guidance from the federal government. Yet, Homeland Security Director Tom Ridge has all he can do simply to catch up to an increasing set of new demands created by September 11, most of them just beginning to take on the basics, such as calibrated alerts and the security safeguards that are necessary to fully reopen National Airport.

Because both the local and federal governments face the same dilemma, it is particularly unjustifiable that nearly all the federal officials invited to testify today have declined. They are the Office of Homeland Security, the Department of Justice, the Federal Bureau of Investigation, the Capitol Police, and the United States Secret Service. Particularly since most of the new security needs or requirements are federal in origin, the least that Congress is entitled to is the kind of testimony that can be presented without harm to national security. When even that testimony is withheld, complaints from administration agencies concerning any legislation that results will be unavailing. I believe that a more cooperative and forthright approach that faces these dilemmas head on, together trying to find solutions, is what is needed.

To help sort out the conundrum of at once opening the society and closing down terrorism, I will shortly be introducing the Open Society with Security Act along with a Senate cosponsor. The bill would authorize a presidential commission to bring together the best minds in the society to investigate how our country can meet the high standards necessary to effectively fight the dangerous menace of international terrorism while accommodating and affirming the central American values of privacy, openness, and public access. Like the Kerner Commission, the Open Society with Security Commission would help us to chart a safe course through deep waters without surrendering the very values that lead us to insist upon defending our country and our way of life.

We can do better than blunt and often untested and ineffective instruments that crush our liberty. I spent my early years at the bar as assistant legal director for the national ACLU, and today I have responsibilities for national security as a member of Congress. My experience in both roles has reenforced my confidence that American ingenuity is ready for the new challenge of winning the struggle against dangerous and dogmatic terrorism and maintaining and enriching the free and open, democratic society that virtually defines our country.

I welcome today's witnesses and appreciate their testimony. I am certain that when we have heard them, all of us will be in a better position to find the appropriate solutions.

Mrs. MORELLA. Thank you, Congresswoman Norton. I am going to ask our first panel to come forward, please. Before they are seated, if you will continue to stand so I can administer the oath that is the tradition before this full committee and subcommittee.

[Witnesses sworn.]

Mrs. MORELLA. The record will reflect affirmative response.

We have for our first panel Kathleen Patterson, chairwoman of the Committee on the Judiciary, the Council of the District of Columbia. Welcome, Ms. Patterson. We also have our police chief, Charles Ramsey, chief of the Metropolitan Police Department of the District of Columbia. And we have John Parsons, associate regional director, National Capital Region of the National Park Service. Margret Nedelkoff Kellems, Deputy Mayor for Public Safety and Justice of the District of Columbia is on her way, and when she arrives, we will swear her in and listen to her testimony. You are all veterans of this subcommittee and other subcommittees and committees, and so I would request that again you keep your comments to about 5 minutes, and the entire testimony that you submit will be included in the record and that will give us a chance to ask this first panel questions. So we start off with you, Chairwoman Patterson.

STATEMENTS OF KATHLEEN PATTERSON, CHAIRMAN, COMMITTEE ON THE JUDICIARY, COUNCIL OF THE DISTRICT OF COLUMBIA; MARGRET NEDELKOFF KELLEMS, DEPUTY MAYOR, PUBLIC SAFETY AND JUSTICE, DISTRICT OF COLUMBIA GOVERNMENT; CHARLES RAMSEY, CHIEF, METROPOLITAN POLICE DEPARTMENT, DISTRICT OF COLUMBIA; AND JOHN PARSONS, ASSOCIATE REGIONAL DIRECTOR, NATIONAL CAPITAL REGION, NATIONAL SERVICE

Ms. PATTERSON. Thank you very much. And good morning. Thank you. As you noted, I am council member Kathy Patterson. I represent Ward 3 and I serve as chair of the council Committee on the Judiciary which has oversight responsibilities for the Metropolitan Police Department and criminal justice issues generally.

Deputy Mayor Kellems and Chief Ramsey will, I am sure, give you an overview of what is in place today in the District in terms of video surveillance employed by the Metropolitan Police Department's Synchronized Operations Command Center, and I am here to share the perspective of the legislature.

First, surveillance, including photo surveillance, is a longstanding and legitimate tool for law enforcement in the District of Columbia, as it is elsewhere. The D.C. Code, for example, includes a definition of the term "law enforcement vehicle" that includes surveillance as a law enforcement activity. The more recent Drug-Related Nuisance Abatement Act of 1999 permits a court to issue an order to abate a nuisance that would include, "use of videotape surveillance of the property and adjacent alleys, sidewalks or parking lots."

And our courts have long had the authority to issue warrants for electronic surveillance in connection with crimes and offenses committed within the District of Columbia.

What is at issue today, as you both have noted, is whether technology itself has blurred or even moved the line between what is longstanding and legitimate law enforcement use of surveillance and what is an unwarranted and potentially illegal violation of privacy rights. Surveillance cameras mounted on District buildings now can monitor the movements of individuals up to and including creating images that can be scanned into a computer and used much as fingerprints are used today.

It is fair to say that are two distinct and divergent reactions to this use of technology: Those who fear a loss of privacy and those who see ready advantage in the use of cutting edge technology as a crime fighting tool. Many District residents have expressed concern about the potential intrusion into their lives and space. Guy Gwynne, on behalf of the Federation of Citizens Associations of the District of Columbia, testified before the Council Committee on the Judiciary in February as follows.

We have Federal and State laws against wiretapping with heavy penalties. Obviously we cannot omit having the same sort of citizen protection extended to the new technology and its largely undefined legal status.

The Federation recommends legislation to define the legal use of surveillance, including protecting privacy and security of file tapes, and civil and criminal penalties for anyone who abuses a surveillance system.

Another witness before the Judiciary Committee in February was Mara Verheyden-Hilliard, who defends a group of demonstrators who have demonstrated in the District of Columbia on international economic policy, and notes that the police department has been surveilling such demonstrations for quite some time, including demonstrations in 2000 and 2001.

At the same council hearing, the ACLU provided research information that I am sure their testimony will provide to this committee today in terms of experiences elsewhere.

I have summarized the viewpoint of the District residents who are concerned about surveillance cameras. Others support the concept wholeheartedly and want to know what they can do to bring cameras into their own neighborhoods. They want surveillance technology as a potential deterrent to neighborhood crime. I would note that 6 years ago the parents of the junior high school my children attended raised their own money to install closed circuit television as a security measure. Since that time D.C. public schools have installed video monitoring systems in 20 elementary schools, 56 junior and senior high schools, with broad support from parents and residents.

The school system is moving forward to having such systems in place in all of the city's schools, paid for in part with Federal Emergency Preparedness funds. The council of the District of Columbia adopted emergency legislation earlier this month to require that guidelines drafted by the police department come to us for review.

This is clearly an issue calling for extensive and wide-reaching public discussion, and I anticipate further public hearings on the issue. The council as a body has not taken a position on the use of surveillance, and I suspect we will, as a group, be of the same two minds I am reflecting in this testimony, concern with privacy rights and concern with use of technology to the maximum extent possible to promote public safety.

Having seen what is available at the police department's operations center, it is easy to understand the potential use of such surveillance; its use, for example, in a city-wide evacuation; its use in responding to a major fire or any other large scale emergency.

The use in protests is another matter. The potential use of image scanning is yet a more difficult issue. I would like to underscore that it is my responsibility and that of my colleagues and Mayor Williams to find the appropriate balance between privacy and public safety and to establish that balance in public policy.

I would also respectfully suggest that this committee and this Congress also needs to take up the issue, and I am delighted to hear Congresswoman Norton mention legislation to provide a national debate on this issue. We all have our work cut out for us in discussing and debating how and whether we use advances in technology, for what purpose and with what result.

Thank you very much and I would be happy to answer questions.

[The prepared statement of Ms. Patterson follows:]

COUNCIL OF THE DISTRICT OF COLUMBIA
WASHINGTON, D.C. 20004

KATHY PATTERSON
COUNCILMEMBER, WARD 3

CHAIRPERSON
COMMITTEE ON THE JUDICIARY

OFFICE: (202) 724-8062
FAX: (202) 724-8118

Testimony by

Councilmember Kathy Patterson

Chair, Committee on the Judiciary, Council of the District of Columbia

Before the

House Committee on Governmental Reform

Subcommittee on the District of Columbia

Hearing on

"Privacy vs. Security: Electronic Surveillance in the Nation's Capital"

March 22, 2002

Thank you for the opportunity to testify today on behalf of the Council of the District of Columbia. I am Councilmember Kathy Patterson. I represent Ward 3 and serve as chair of the Council Committee on the Judiciary, which has oversight responsibility for the Metropolitan Police Department and criminal justice issues generally.

Deputy Mayor Kellems and Police Chief Ramsey will, I am sure, give you an overview of what is in place today in the District of Columbia in terms of the photo surveillance technology employed at the Metropolitan Police Department Synchronized Operations Command Center. I would like to share the perspective of the legislature.

First, surveillance including photo surveillance is a longstanding and legitimate tool for law enforcement in the District of Columbia, as it is elsewhere. The D.C. Code, for example, includes a definition of the term "law enforcement vehicle" that includes surveillance as a law enforcement activity (Section 50-702 in the Alternative Fuels Technology Amendment Act of 1993). The Council-passed Drug Related Nuisance Abatement Act of 1999 permits the Court to issue an order to abate a nuisance that would include "use of videotaped surveillance of the property and adjacent alleys, sidewalks or parking lots." And our Courts have long had the authority to issue warrants for "electronic surveillance in connection with crimes and offenses committed within the District of Columbia" (Section 11-941, effective July 29, 1970).

What is at issue today is whether technology itself has blurred or moved the line between what is longstanding and legitimate law enforcement use of surveillance, and what is an unwarranted and potentially illegal violation of privacy rights. Surveillance cameras mounted on District buildings now could monitor the movements of individuals up to and including creating images that can be scanned into a computer and used much as fingerprints are used today.

It is fair to say there are two distinct and divergent reactions to this use of technology: those who fear a loss of privacy and those who see ready advantage in the use of cutting edge technology as a crime-fighting tool.

First, many District residents have been horrified at the potential intrusion into their lives and space. Guy Gwynne, on behalf of the Federation of Citizens

Associations of the District of Columbia, testified before the Council Judiciary Committee on February 25, as follows: "We have federal and state laws against wiretapping, with heavy penalties. Obviously we cannot omit having the same sort of citizen protection extended to the new technology and its largely undefined legal status. Any new legislation must start from the premise: Is broad continuous surveillance necessary, and, also, is it legally defensible? Video surveillance can easily become an intrusive search without a warrant and without probable cause or individualized suspicion."

The Federation is recommending legislation to define the legal use of video surveillance, including protections of privacy and security of file tapes, to provide for swift destruction of tapes without value as evidence, and civil and criminal penalties for anyone who abuses a surveillance system. Mr. Gwynne concluded: "In the absence of such legislation, the public cannot systematically be protected from abuses of privacy. The use of hidden cameras could undermine people's expectation of anonymity in their private affairs. Justice Louis Brandeis's thought still holds: the most precious right of civilized people is the right to be left alone."

Another witness before the Judiciary Committee on February 25 was Mara Verheyden-Hilliard, cofounder and attorney with the Partnership for Civil Justice and lead attorney in a case brought against the District by individuals protesting international economic policy. She asserted that the Metropolitan Police Department has been "surveilling law abiding political protestors for nearly two years now, well before September 11" including the demonstrations during the International Monetary Fund meetings in 2000 and 2001.

Ms. Verheyden-Hilliard asked: "Why is it that we assume that the hundreds of thousands of residents of this city, and hundreds of thousands of people who come to this city to work or as tourists, law abiding people, should be subject to extreme surveillance and tracking without probable cause, without even reasonable suspicion?" She asks, further: "Are we going to willingly give DC and federal law enforcement the ability to track our movements on foot and in our cars, log our associations with other people, read our correspondence over our shoulders?"

At the same Council hearing the ACLU of the National Capital Area summarized experiences in other jurisdictions with camera surveillance. The Oakland, California police department used surveillance cameras in public places for three years but concluded that "there is no conclusive way to establish that the presence of video cameras resulted in the prevention or reduction of crime." Detroit, said the ACLU, took 14 years before deciding to abandon its surveillance camera system in 1994. And Tampa, Florida, used facial recognition technology for several months and abandoned the system last summer. Said the ACLU: "It never identified a single face in its database of suspects."

I have here summarized the viewpoint of many District residents who were very concerned at the notion of surveillance cameras mounted on buildings around the city. Other residents support the concept and simply want to know what steps they can take to ask to have surveillance cameras mounted in their neighborhood -- on their block and on their corner. They desperately want the surveillance technology as a potential deterrent to neighborhood crime.

In 1996 parents at the junior high school my children attended, Alice Deal Junior High School, raised money to install a closed-circuit television system as a security measure in what is the city's largest junior high school. To date, the D.C. Public Schools have installed video monitoring systems in 20 elementary schools and 56 junior and senior high schools, with broad support from parents and residents. According to school system officials, they are moving forward to have monitoring systems in place in all of the 163 schools and other sites, paid for in part with federal emergency preparedness funds.

The Georgetown Business Association paid for surveillance camera technology at Wisconsin and M Streets for use during special events in that neighborhood. My colleague, Councilmember Sharon Ambrose, takes the view that the new guidelines Chief Ramsey and his staff are drafting for MPD's use of surveillance technology should include a public procedure a neighborhood could follow to secure exactly the same technology other D.C. residents fear.

The Council of the District of Columbia adopted emergency legislation earlier this month to require that the guidelines drafted by the MPD come before the Council for review. That resolution reflected the divergent viewpoints of our citizens, stating, "Maintaining a balance between the need for effective law enforcement and the need to protect the privacy of the citizens of and visitors to the District of Columbia is an important public policy issue for the District's elected leaders to consider." This clearly is an issue calling for extensive and wide reaching public discussion, and an issue for elected leaders to review. I anticipate that the Council will hold a public hearing on the specific guidelines drafted by the MPD.

The Council as a body has taken no position on the use of video surveillance, and I suspect we reflect the same divergent opinion as our constituents. I myself agree both with those concerned about privacy rights and those who wish to use technology to the maximum extent possible to promote public safety. Having seen the pictures available at the SOCC, it is easy to understand the potential use of video surveillance in a District-wide evacuation; in a major fire, or another large-scale emergency. The use in protests is another issue; the potential use of image scanning is another more difficult issue. It is my responsibility, and that of my colleagues and Mayor Williams, to find the appropriate balance between privacy and public safety, and establish that balance in public policy.

Having noted that the Council will review the issue of video surveillance in the near future, I would also respectfully suggest to this Committee that if the Congress is not already looking at the issue of new technology and its impact on public policy, that it do so. How the nation uses technological advances is not just a local issue. Our use of technology has ramifications for intelligence, anti-terrorism planning, and law enforcement at the national level, and it has clear potential impact on continued protection of Constitutional rights.

To illustrate the national nature of the issue, I am sure you are aware of the Supreme Court decision last summer in which the high court came down on the side of privacy rights against law enforcement surveillance – but on a 5-4 vote. In *Kyllo vs. United States* the high court overturned a decision by the 9th Circuit in San Francisco and ruled that police use of a thermal imaging device to detect patterns of heat coming from a private home is a search that requires a warrant. Justice John Paul Stevens dissented: he held that that the police activity in the Oregon case "did not invade any constitutionally protected interest in privacy." This is a critical issue: and a fascinating one when you consider how the traditional voting patterns on the court shifted on this particular case.

We all have our work cut out for us in discussing and debating how – and whether – we use advances in technology, for what purpose, and with what result on behalf of residents.

Thank you, and I would be happy to answer any questions.

Mrs. MORELLA. Thank you very much, Chairwoman Patterson. And you have been concise, and I note that you have extrapolated from your submitted written testimony. Pleasure to have you with us as always.

Chief Ramsey. Thank you, sir.

Mr. RAMSEY. Good morning, Madam Chair, Congresswoman Norton, members of the subcommittee staff and guests.

In recent weeks an awful lot has been written about the Metropolitan Police Department's use of closed circuit TV or CCTV. Much of the reporting has been factual. Regrettably, some of it has been less than accurate, and I applaud the subcommittee for calling today's hearing, and I thank you for the opportunity to inject not only the facts, but also some perspective into this discussion.

Let me state up front that the Metropolitan Police Department welcomes public scrutiny for our policies and programs. Our joint operations command center remains an open book, accessible to news reporters, law enforcement officials and political leaders from across the country and around the world.

When the ACLU and others expressed concern about our use of CCTV, we invited those groups in for a demonstration and meeting. The bottom line: We welcome public debate, but we ask that the debate be based on facts, not on conjecture.

Fact No. 1. The Metropolitan Police Department is using CCTV in public spaces only in a limited legal and responsible manner. We are not running a 24-hour-a-day, 7-day-a-week video monitoring operation. The joint operations command center is activated only during major events in our city: Large demonstrations, Presidential inaugurations and the like, or during periods of heightened alert for terrorism such as the weeks immediately following September 11th and during the Olympic Games in Salt Lake City.

During other times, including right now, the center is not operational. Our use of CCTV is need-driven, when there are specific threats and tangible public safety benefits from activating it.

Another misconception is that our department has a vast network of hundreds of cameras at our disposal at any moment. We do not. Our current system includes approximately a dozen cameras, mounted on buildings in downtown D.C. and focused on high-risk targets for terrorism, including the National Mall and public spaces outside Union Station, the White House and the Capitol.

During times of heightened alert, the cameras gave us a clear real-time view of those potential targets without having to dedicate police officers on the ground to this type of monitoring activity. Our use of CCTV is legal. The cameras monitor only public spaces. There is no audio overhear capability in our system. As such, we are not engaging in electronic surveillance as defined in law.

And, finally, the use of CCTV is sensitive to the privacy expectations of individuals. When our cameras are operational, they generally focus on broad public spaces, not on individuals within those areas, and we do not employ any type of face recognition technology.

Fact No. 2. The Metropolitan Police Department is working to link our joint operations command center with other public agency closed-circuit television systems that monitor public spaces, but we have no interest in linking with privately operated networks that monitor private space.

We are particularly interested in linking with the traffic cameras operated by transportation agencies in D.C., Maryland and Virginia, giving us real-time information on traffic flow and bottlenecks during major events or evacuations. We have already successfully tested a linkage with the CCTV system operated by the D.C. Public Schools. This type of real-time visual information would be extremely valuable in responding to a Columbine-like incident. And we have begun discussions with Metro about linking with its video system in stations and platforms.

In all of these instances, there is one critically important safeguard: Access to these systems is controlled by the agency that operates the cameras, not by the Metropolitan Police Department.

With the public schools, for example, there are strict protocols governing access to their CCTV system. Only the schools, not the police department, can activate the system, and only the schools can allow us in.

With respect to privately operated video networks, the Metropolitan Police Department is not linking with such systems now, nor do we have any plans to do so. We have absolutely no interest in peering into the private activities of anyone.

Fact No. 3. The Metropolitan Police Department is very carefully evaluating any expansion into neighborhood-based cameras. Currently we are able to access a camera mounted near the corner of Wisconsin Avenue and M Street Northwest in the Georgetown commercial-entertainment district. That camera was purchased by the business and professional community which then approached our department about monitoring the public space images from a community policing center in Georgetown. Our department agreed to this request in part because we want to evaluate the operational feasibility and public safety benefits of a neighborhood-based camera.

In recent weeks our department has received several inquiries from our community groups requesting CCTV in their neighborhoods to help combat robberies, thefts and other street crime. We are carefully studying the issues involved, including cost, scope and length of operation, resources needed to monitor the cameras, and privacy issues.

With any neighborhood-based installations, there are certainly principles that we would follow. First, any cameras would have to target a specific crime problem for which CCTV technology may be beneficial.

Second, there would have to be extensive dialog with the community about the deployment of cameras and widespread community support for their use.

Fact four. As technology has advanced, our policies and procedures have not always kept pace. One of the concerns raised by ACLU and others, and shared by myself and other MPD leaders, was that policies and procedures governing our use of video were not as specific and formalized as they should be. Who could activate the command center, who controls the cameras, what images would be recorded, if so, how long would the tapes be retained?

These are legitimate issues that need to be clarified. Over the last month our department has initiated a fast track process to develop now policies in this area. We are currently finalizing a new department directive for the Mayor's review and approval on the use of CCTV. The development of stronger policies and procedures will not only enhance public confidence in the system, but also safeguard against any possible abuses.

And, finally, the Metropolitan Police Department does not view CCTV as a panacea in achieving either neighborhood safety or homeland security. It is just one more tool that can support these efforts. Our intention is not to somehow transform policing in our city from a neighborhood-based community policing strategy into one that hinges on CCTV.

This technology, when used properly, can support community policing. But I know it will never replace community policing. We are in a unique time, in a unique city, a city that faces not only serious crime problems, but also the very real threat of terrorism. And the Metropolitan Police Department has the unique responsibility of protecting not only our residents, but also the millions of people who come to our Nation's Capital every year to work, to visit, to petition their government.

Given the enormous challenges we face, I would argue that it would be irresponsible for us not to use every legal tool at our disposal, including video technology, to help protect our city, and ultimately our democracy. We will continue to use these tools judiciously, responsibly, openly, and with strong public oversight. Thank you.

[The prepared statement of Mr. Ramsey follows:]

Hearing on Privacy vs. Security:
Electronic Surveillance in the Nation's Capital

United States House of Representatives
Committee on Government Reform
Subcommittee on the District of Columbia

The Honorable Constance A. Morella
Chairwoman

Charles H. Ramsey
Chief of Police
Metropolitan Police Department

March 22, 2002

Madame Chair, Congresswoman Norton, other members of the Subcommittee, staff, and guests: In recent weeks, an awful lot has been written and spoken about the Metropolitan Police Department's Joint Operations Command Center and our use of video technology within that center. Much of the reporting and analysis has been factual and reasonably accurate. Regrettably, some of it has been less than accurate – and some has been pure fantasy. That is why I applaud the Subcommittee for calling today's hearing, and I thank you for the opportunity to inject not only facts, but also some perspective, into this discussion.

Let me state, up front, that the Metropolitan Police Department welcomes public scrutiny of, and public debate over, our policies, programs and actions. Our Joint Operations Command Center, and the technology behind it, have been – and remain – an open book. We have made the JOCC accessible to news reporters from throughout the region, across the country and around the world. We have demonstrated the center to law enforcement and political leaders from Europe, Asia, and North and South America. (Let me add that our open invitation to members of the Subcommittee and to staff remains, at your earliest convenience.) And when the ACLU and other privacy rights groups expressed concern about our use of video technology, we immediately invited representatives of those groups in, for a demonstration and an open and frank airing of their concerns. The bottom line: we welcome public debate on this issue, but we ask that the debate be based on facts, not on conjecture or conspiracy theories.

Fact #1: The Metropolitan Police Department is using video surveillance of public spaces only in a limited, legal and responsible manner.

Perhaps the biggest misconception in this whole area is that we are operating some type of 24-hour-a-day, 7-day-a-week video monitoring operation. We are not. The Joint Operations Command Center is activated only during major events in our city – marches or demonstrations, Presidential Inaugurations and the like – or during periods of heightened alert for terrorism. In fact, the JOCC was scheduled to be activated for the first time during the September 2001 meetings of the IMF and World Bank Group, but the center was pressed into action ahead of schedule on September 11th. The JOCC was up and operational prior to the jet slamming into the Pentagon that morning, and it provided critical law enforcement support in the days and weeks that followed. Understand that the JOCC is not a video monitoring center, and that video is only one element of a multi-agency command complex that has become critically important in the post-9/11 environment.

The center was brought down several weeks after September 11th, but was re-activated on February 11, following the specific terrorist alert related to February 12th. The center remained in operation until the end of the Olympic Games in Salt Lake City on February 24th; other than a few demonstrations, it has not been operational since then. My point is that our use of video is need-driven – when there are specific and tangible public safety benefits from using the technology. Put another way, we are using video because we "should," not simply because we "can."

Another misconception is that our Department somehow has a vast network of hundreds of cameras at our disposal at any moment. We do not. Our current system has approximately one dozen cameras that are mounted on buildings in downtown DC and focused on critical areas that are at high risk of terrorist attacks. These sites include the National Mall, monuments and museums, the plaza outside Union Station, the public areas surrounding the White House and the Capitol, as well as some of the major highways leading into the downtown area. Geographically, these areas represent only a small sliver of Washington, DC, but we believe they represent the highest risk targets when it comes to terrorism. During times of heightened alert, the video cameras give us a clear, real-time view of these potential targets, without having to dedicate police officers on the ground to this type of monitoring activity. We can also accept video feeds in our Command Center from both the U.S. Park Police helicopter and the MPD's Falcon One helicopter.

In addition to being limited in scope, our use of video is also legally grounded in both the Constitution and case law. The cameras monitor only public spaces, and there is absolutely no audio overhear capability in our system. "Electronic surveillance," as defined by law, typically involves audio overhear, which does require judicial approval. Our video-only system of monitoring public space is not "electronic surveillance" in this legal sense.

Finally, our use of video is responsible and sensitive to the privacy rights and expectations of individuals. When our cameras are operational, they focus on broad public areas, not on the individuals within those areas. For example, if individual members of the Subcommittee were to walk across Columbus Circle when our camera there was activated, people in the Command Center would be hard pressed to distinguish Madame Chair from Mrs. Norton from anyone else in the picture. Understand that these are wide-angle shots intended to pick up suspicious activities, not to track specific individuals. Of course, were someone to suddenly fall down clutching his chest and a crowd gathered, technicians could zoom-in the camera for a closer look – and potentially speed up the dispatch of a life-saving ambulance. Similarly, if officers were to detect a large truck stopping in a "No Parking" zone, with the occupants jumping out of the vehicle and running, the camera could zoom-in on the suspicious truck and on the suspicious individuals. That real-time information could potentially thwart a catastrophe and help apprehend potential terrorists. I am not suggesting that this or similar scenarios have happened, but in light of the events of September 11th, they are certainly not implausible. The availability of video would enable us to detect potential threats more quickly and to respond more effectively and safely.

Fact #2: The Metropolitan Police Department is working to link our Joint Operations Command Center with other public agency video networks that monitor public spaces. But we are not currently linking with, nor do we plan to link with, privately operated networks that monitor private space.

In addition to the dozen cameras owned and operated by the MPD, our Department is testing the ability to link with existing public agency video networks. We are particularly interested in tying into traffic camera systems operated by transportation agencies in DC, Maryland and Virginia. The benefits of being able to see, in real time, traffic flow and bottlenecks during major events or evacuations are obvious. We have already successfully tested a linkage with the video system operated by the DC Public Schools inside its high schools. As part of a mock exercise of our response to a Columbine-like incident inside one of DC's schools, we were able to project video feeds from inside the test school into our JOCC. Were that scenario to be real, the real-time video information would have been invaluable in allowing us to assess the situation thoroughly and accurately, to identify what resources were needed, and to determine points of entry – all without unduly endangering innocent students, faculty, staff or our officers. More recently, we have initiated discussions with the Metro system about developing and testing similar linkages with its video system in underground stations and platform.

In these and any other linkages we may develop with publicly operated video systems, there is one critically important safeguard that people need to understand. Access to these outside systems is controlled by the agency that operates the cameras, not by the Metropolitan Police Department. The only way the MPD will be able to access the DCPS video system is if school officials ask us to do so, and then provide the "key" to grant us access. Technologically and operationally, it is impossible for the MPD to simply "flick a switch" and take a look at what is happening inside Ballou or Dunbar or Wilson High School this morning. More importantly, absent a specific request from the schools based on a legitimate public safety need, the MPD would not want to engage in any such "fishing" expedition inside our schools, our subway system or other public facilities.

With respect to privately operated video networks, the MPD is not now linking with such systems, nor do we have plans to do so. Private cameras in the lobbies and elevators of apartment buildings and hotels will not join our network, as one national columnist recently stated (William Safire, "The Great Unwatched," The New York Times, February 18, 2002). Nor do we have any interest whatsoever in accessing video from hospital and nursing home rooms, as that same columnist suggested. And the 35-millimeter still-camera images that are taken with our red-light and photo radar cameras are not only technically incompatible with our video network, but of extremely limited value from a law enforcement or homeland security perspective. We are not working to link those cameras either.

Our mission is to protect the residents, workers and visitors in the District of Columbia. To the extent that limited video monitoring of public space allows us to achieve that goal, without interfering with personal privacy rights, we will engage in such monitoring. But we have absolutely no interest in peering into the private activities of anyone.

Fact #3: The Metropolitan Police Department is very carefully and cautiously evaluating any expansion of the current, limited network into neighborhood-based applications.

Our use of video technology thus far has been limited to the downtown DC cameras I mentioned earlier, as well as one camera mounted near the corner of Wisconsin Avenue and M Street, NW, in the Georgetown commercial and entertainment area. That camera was purchased by the business and professional community in Georgetown, which then approached our Department about monitoring the images on an as-needed basis, from a community policing center located in the Georgetown community. Our Department agreed to this request, in part because we wanted to evaluate the operational feasibility and public safety benefits of neighborhood-based cameras.

In recent weeks, our Department has received several requests from community groups and individual residents seeking video cameras in their neighborhoods. Many of these people are frustrated with the level of crime in their communities – especially drug dealing, robbery, auto theft, theft from auto and other street crimes. And they believe that police- or even community-operated video cameras could provide both deterrence and enforcement benefits. Our Department cannot ignore these community requests, and we are studying the issues surrounding neighborhood-based cameras very carefully. Some of the issues we must consider are costs, scope and length of operation, and the resources needed to monitor the cameras when they are operational.

With any future neighborhood-based installations, there are certain principles we will follow. First, the cameras would have to target a specific public safety problem for which video technology may prove beneficial. Second, there would have to be extensive and ongoing dialogue with the community about the deployment and operation of the cameras. Finally, there would have to be widespread community support for the use of the technology. The Metropolitan Police Department has absolutely no interest in putting cameras in communities that do not want them or for purposes that do not have clear public safety benefits.

Fact #4: As technology has advanced, policies and procedures on its use have not always kept pace. The Metropolitan Police Department recognizes this shortcoming, and we are working aggressively to develop and implement policies and procedures that address the key issues raised by this technology.

One of the concerns aired by the ACLU and others during our meeting last month was that policies, procedures and guidelines governing our use of video were not as specific and formalized as they should be. In particular, these groups expressed concern about who could activate the Command Center, who controlled the cameras, would the images be recorded, and if so, how long would the recordings be retained, among other matters. All of these are legitimate issues that need to be clarified, and our Department has nearly completed a fast-track process to develop and implement new policies in this area. As part of this process, we invited the ACLU and others to provide information and input on the front end, and the opportunity to review our drafts on the back end. I respect the perspectives and insights of the individuals we met with, and our Department is committed to addressing their concerns to the extent possible. We are currently putting the finishing touches on the final draft of a new Department directive, for the Mayor's review and approval, on the use of video technology, and we will certainly share it with the Subcommittee once it is finalized.

The development and implementation of these new policies and procedures will not only enhance public confidence in the system, but also improve our operational effectiveness. In particular, they will help us safeguard against any abuses of the system. This is critically important. For although I am confident that we have used video technology in a legal, ethical and responsible manner thus far, I am fully aware that some police officers in the past have abused their authority by using technology to harass, intimidate and even blackmail citizens. I reject the proposition that, just because the police have the technology, they will abuse it. But I certainly understand the serious concerns that some individuals and groups have expressed, based on their past experiences. I want to assure the Subcommittee that misuse of this technology will not be tolerated on my watch. The promulgation of these new policies and procedures will help ensure that is the case.

Fact #5: The Metropolitan Police Department does not view video technology as a panacea in achieving either neighborhood safety or homeland security. It is simply one additional tool that can certainly support these efforts.

As potentially useful as video may be in certain circumstances, it is not a panacea, and the Metropolitan Police Department is not approaching it as such. While our Department and our city can learn a lot from the experiences of Great Britain, Australia and elsewhere, learning from them does not mean we are trying to emulate them.

I am committed to having as many of our uniformed police officers as possible out on the street, fighting crime and partnering with the community, not sitting behind a video monitor. Of course, if the targeted use of video can help the "cop on the beat" by providing him or her with an extra set of eyes in targeting a chronic problem, then we will look to provide the officer with that tool. But our intention is not to somehow transform policing in our city from a neighborhood-based, community policing strategy into one that hinges on video surveillance. I believe that, when used properly, video technology can help support community policing. But I know that it will never replace community policing.

We are in a unique time, and in a unique city – a city that faces not only serious crime problems, but also the very real threat of terrorism. In this unique time and in this unique city, the Metropolitan Police Department has the unique responsibility of protecting not only our residents, but also the millions of people who come to our Nation's Capitol every year to work, to visit, to petition their government.

Given this uncertain environment and the enormous challenges we face, I would argue that it would be irresponsible for us not to use every legal tool at our disposal, including video technology, to help protect our city and, ultimately, our democracy. We will continue to use these tools judiciously, responsibly, openly and with strong public oversight. Thank you very much.

Mrs. MORELLA. Thank you, Chief Ramsey. John Parsons, the Associate Regional Director of the National Capital Region of the National Park Service. And then we will give you a chance to catch your breath, Ms. Kellems. Thank you.

Mr. PARSONS. Good morning, Madam Chairwoman and Mrs. Norton. I appreciate the opportunity to present the views of the Department of Interior on the issue of privacy and security with respect to the use of electronic surveillance in the Nation's Capital.

The National Park Service is privileged to have the responsibility for managing some of our Nation's most treasured symbols, including ones in the monumental core of Washington, D. C. The monumental core includes the Washington Monument, the Lincoln, Jefferson and Franklin Delano Roosevelt Memorials, the White House complex, and the Korean War Veterans and Vietnam Veterans War Memorials.

For several years the National Park Service, with the guidance of the U.S. Park Police, our urban law enforcement arm, has been working on enhancing security in and around the heavily visited monumental core. Numerous studies have been conducted by the National Park Service and its consultants in recent years to assess potential threats to the sites in the monumental core, the level of security present and actions that should be taken to increase security and thus minimize the risk of danger to those who visit these structures.

One recommendation, which is common to nearly all of the studies, is the use of closed-circuit TV, a 1999 study of the Booz-Allen-Hamilton which focused on terrorist threats and recommended the installation of CCTV in all of the monuments and memorials within the monumental core.

Although the process of planning the CCTV system and obtaining funding for it had begun prior to September 11th, the installation of this technology became a higher priority after the tragic events of that date.

Installation of CCTV at sites in the monumental core is part of a larger effort to increase security at National Park Service sites that may be at high risk for terrorist activity for which the administration is seeking substantial increases in the fiscal year 2003 budget.

The budget requests include an increase of $12.6 million for the U.S. Park Police in Washington, DC, and New York City, to fund additional recruitment classes, equipment and overtime.

It also includes approximately $13 million at the Washington Monument, $6.2 million at the Lincoln Memorial and $4.7 million at the Jefferson Memorial for vehicle barriers, security lighting, and associated improvements. And it includes an increase of $6.1 million for increased security at park units across the country that are national icons as well, such as the Statue of Liberty, Independence Hall, the Arch of Western Expansion in St. Louis, and Mt. Rushmore.

The National Park Service is not currently using any CCTV in the monumental core area. However, within the next 6 months we plan to have CCTV installed at six sites, the Washington Monument, the Lincoln Memorial, the Jefferson Memorial, the Vietnam

Veterans Memorial, the Korean War Veterans Memorial and the FDR Memorial.

Park Police personnel will continuously monitor these cameras as a Park Police facility. The current plan calls for the images to be recorded on a continuous loop which will record over itself after a yet-to-be determined period of time.

Recording the images will allow the police to save any that are needed for evidentiary purposes. The estimated cost of this system is approximately $2 to $3 million. The National Park Service plans to use cameras monitored by the U.S. Park Police only in public areas where there is no expectation of privacy. The images that are recorded would be used only for valid law enforcement purposes. At this time the National Park Service is planning to install CCTV only in the monumental core area. We do not have any plans to use any other type of surveillance technology such as facial recognition types of CCTV.

The U.S. Park Police operate in New York City and San Francisco as well as Washington, DC. In New York, the Park Police have cameras in place on Liberty Island and in the Statue of Liberty that are monitored on a continuous basis with a loop recording system. Park Police personnel in the New York field office are working on plans to upgrade the entire security system in the vicinity of the Statue of Liberty, including using digital CCTV.

In San Francisco, the Park Police have not installed any cameras and do not have current plans to do so. However, the Golden Gate Bridge Authority utilizes numerous cameras on that facility which is directly adjacent to our park.

In summary, we see CCTV as, used appropriately, as a cost-effective and nonintrusive way to monitor and protect larger areas than would able to be protected with available personnel. It is thus an important tool that can help the National Park Service safeguard the national treasures under our stewardship and the people who visit them.

Madam Chairwoman, that concludes my testimony. I would be happy to answer any questions I may have stimulated. Thank you.

[The prepared statement of Mr. Parsons follows:]

STATEMENT OF JOHN PARSONS, ASSOCIATE REGIONAL DIRECTOR FOR LANDS, RESOURCES, AND PLANNING, NATIONAL CAPITAL REGION, NATIONAL PARK SERVICE, DEPARTMENT OF THE INTERIOR, BEFORE THE SUBCOMMITTEE ON THE DISTRICT OF COLUMBIA OF THE HOUSE COMMITTEE ON GOVERNMENT REFORM, REGARDING ELECTRONIC SURVEILLANCE IN THE NATION'S CAPITAL

March 22, 2002

Madam Chairwoman, thank you for the opportunity to present the views of the Department of the Interior on the issues of privacy and security with respect to the use of electronic surveillance in the Nation's Capital.

The National Park Service is privileged to have the responsibility for managing some of our Nation's most treasured symbols, including ones in the Monumental Core of Washington, D.C. The Monumental Core includes the Washington Monument; the Lincoln, Jefferson, and Franklin Delano Roosevelt Memorials; the White House complex; and the Korean War Veterans and Vietnam Veterans Memorials. For several years, the National Park Service, with the guidance of the U.S. Park Police, our urban law enforcement arm, has been working on enhancing security in and around the heavily visited Monumental Core.

Numerous studies have been conducted by the National Park Service and its consultants in recent years to assess potential threats to sites in the Monumental Core, the level of security present, and the actions that should be taken to increase security and thus minimize the risk of danger to those who visit and work at these structures. One recommendation, which is common to nearly all of the studies, is the use of closed circuit television (CCTV). A 1999 study by the firm Booz-Allen & Hamilton, which focused on terrorist threats, recommended the installation of CCTV in all of the monuments and memorials within the Monumental Core. Although the process of planning the CCTV system and obtaining funding for it had begun prior to September 11, the installation of this technology became a higher priority after the tragic events of that date.

Installation of CCTV at sites in the Monumental Core is part of a larger effort to increase security at National Park Service sites that may be at high risk for terrorist activity, for which the Administration is seeking substantial increases in the Fiscal Year 2003 budget. The budget request includes an increase of $12.6 million for the U.S. Park Police in Washington, D.C. and New York City to fund additional recruitment classes, equipment, and overtime. It also includes approximately $13 million at the Washington Monument, $6.2 million at the Lincoln Memorial, and $4.7 million at the Jefferson Memorial for vehicle barriers, security lighting, and associated improvements. And, it includes an increase of $6.1 million for increased security at park units across the country that are National icons, such as the Statue of Liberty, Independence Hall, the Arch of Western Expansion in St. Louis, and Mount Rushmore.

The National Park Service is not currently using any CCTV in the Monumental Core area. Within the next six months, the National Park Service plans to have CCTV installed at six sites: the Washington Monument, the Lincoln Memorial, the Jefferson Memorial, the Vietnam Veterans Memorial, the Korean War Veterans Memorial, and the Franklin Delano Roosevelt Memorial. Park Police personnel will continuously monitor these cameras at a Park Police facility. The current plan calls for the images to be recorded on a continuous loop, which will record over itself after a yet-to-be-determined period of time. Recording the images will allow the police to save any that are needed for evidentiary purposes. The estimated cost for this system is approximately $2-3 million.

The National Park Service plans to use cameras monitored by the U.S. Park Police only in public areas where there is no expectation of privacy. The images that are recorded would only be kept for a certain amount of time, as noted above, and would be used only for valid law enforcement purposes. At this time the National Park Service is planning to install CCTV only in the Monumental Core area. We do not have any plans to use any other type of surveillance technology, such as facial recognition types of CCTV.

The National Park Service had some discussions with the Metropolitan Police Department (MPD) last year concerning the installation of cameras within the Monumental Core area that would be in addition to the ones the National Park Service already plans to install. MPD has concurrent jurisdiction over the Monumental Core and other areas in the District of Columbia that are under the primary jurisdiction of the National Park Service. Since the U.S. Park Police already had plans in place for cameras at these locations, the National Park Service declined the MPD offer. The National Park Service is aware of plans by the Department of PublicWorks to install cameras for traffic monitoring purposes, and is in discussion with that agency about the possibility of placement in areas that are under the jurisdiction of the National Park Service, such as Rock Creek Park.

The U.S. Park Police operate in New York City and San Francisco as well as Washington, D.C. In New York, the Park Police have cameras in place on Liberty Island and in the Statue of Liberty that are monitored on a continuous basis with a loop recording system. Park Police personnel in the New York Field Office are working on plans to upgrade the entire security system in the vicinity of the Statue of Liberty, including using digital CCTV. In San Francisco, the Park Police have not installed any cameras and do not have currently have plans to do so. However, the Golden Gate Bridge Authority utilizes numerous cameras on that facility, which is adjacent to our park areas.

In summary, we see CCTV, used appropriately, as a cost-effective, non-intrusive way to monitor and protect larger areas than we would be able to protect with available personnel. It is thus an important tool that can help the National Park Service safeguard the National treasures under our stewardship and the people who visit them.

Madam Chairwoman, that concludes my prepared remarks. I would be happy to answer any questions you may have.

Mrs. MORELLA. Thank you, Mr. Parsons. We will get to the questions in a few moments, but right now I would like to swear in Margret Nedelkoff Kellems, the Deputy Mayor of Public Safety and Justice. If you would stand and raise your right hand, Ms. Kellems.

[Witness sworn.]

Mrs. MORELLA. Thank you. Affirmative response. And now we would love to hear your comments.

Ms Kellems. Good morning, Chairwoman Morella and Congresswoman Norton. Please accept my apologies for being late this morning. I am Margret Nedelkoff Kellems, the Deputy Mayor for Public Safety and Justice. I thank you for convening this hearing on a very high profile and often misunderstood topic of the District's use of CCTV as a tool for ensuring the public safety.

I would like to take just a few minutes to outline the Mayor's priorities and concerns regarding the use of this powerful tool, and what steps we are taking to ensure that our practices are judicious and relative to our needs.

Chief Ramsey has already discussed the specifics of the current operation of the system and the policies and procedures that MPD is drafting to govern its operation. Mayor Williams has consistently identified neighborhood safety and qualify of life among his top priorities for his administration.

Since the events of September 11th, homeland or, in this case, hometown security has taken on a new significance, particularly in our Nation's Capital. Mayor Williams continues to stress that in our pursuit of these goals, we must take great pains not to unduly impact the personal privacy interests of our citizens, the many people who work in our city, and the more than 20 million visitors who visit annually.

CCTV is not a new law enforcement tool. In fact, law enforcement has been using CCTV for years as a means of collecting criminal evidence. As you may well imagine, without CCTV, MPD would not have nearly the success it has had in closing down many of the drug markets that impact our city.

Of course, what we are talking about today is substantially different than using video technology in targeted law enforcement efforts, and we appreciate the opportunity to discuss this issue in an open forum with you today.

As CCTV has moved to the forefront as an effective tool in ensuring the public safety and protecting against threats to our city, Mayor Williams has committed to an open dialog and discussion of the benefits and concerns on both sides of the issue. We recognize that engaging all of the stakeholders in the development of our system, both the security experts and those seeking to protect our privacy rights, is the only way we will come to an acceptable result that everyone can live with.

The Mayor has been very clear, and all of the stakeholders agree, that the primary objective of this system is to enhance public safety during major events, times of heightened alert, and actual emergencies, whether or not they are terrorism-related.

During these times, law enforcement resources are our most valuable commodity in terms of ensuring safety and peace. CCTV is tremendously useful in helping us allocate and manage those resources effectively. Instead of relying on radio call-in information from officers scattered around the event or the city, officers whose field of vision is limited to their immediate surroundings, CCTV allows us to monitor large and distant areas quickly and unobtrusively. That information allows us to redirect officers, know where reinforcement is needed, and anticipate where we might need other types of equipment or response.

This information assists in protecting the public as well as our first responders. In fact, the utility of CCTV at major events was proven before September 11th. MPD leased video technology equipment during the IMF World Bank demonstrations, the inaugural, and even the NBA All Star game to assist in resource deployment during those events.

Based on those successes and in anticipation of the planned IMF World Bank meetings and demonstrations scheduled for the end of September last year, MPD began the development of a small video network capability in its joint operations command center.

Since September 11th, MPD has expanded the video network to approximately one dozen cameras focused on areas of potential terrorist threats, and has been pursuing linkages with the video systems operated by area transportation departments and Metro.

Even while we pursue the use of CCTV in these kinds of situations, Mayor Williams has asked MPD to explore the possible future uses of technology for controlling crime on a daily basis. Of course, that is a big step from where we are now, and there are many, many issues that must be considered and evaluated before we move forward. Questions regarding individual privacy rights as well as important operational concerns like the location of the cameras, how they are monitored and by whom and from where, are the video feeds tape-recorded for evidence, and how long and for what purposes are those tapes maintained, would have to be worked out in great detail, and MPD is already engaged in that process.

It is important to note that CCTV is not intended to be a primary neighborhood policing tool. There is no substitute for the security offered by the presence of officers patrolling the neighborhoods themselves. That said, we must be responsive to the public concerns about the potential uses of CCTV once that capacity is in place.

Several neighborhoods have requested that cameras be installed in their communities, hoping that it would have a deterrent effect on criminal activity, and perhaps even assist in capturing and prosecuting criminals. Certainly there will be disagreement among different communities regarding the appropriateness of CCTV.

Bear in mind that at the core of MPD's policing philosophy is respect for the interest of its communities. Chief Ramsey's Partnerships in Problem Solving model has as its core value that police are to work with the communities to solve problems together in a way that is acceptable to all involved.

But the fact remains that in some places, in certain communities, stores or at ATM machines and around government buildings, for example, video monitoring equipment has helped control crime and convict criminals. Mayor Williams has instructed us to assess the benefits and burdens of these implications to determine whether, on balance, CCTV as a community crime fighting tool is effective, even as we move increasingly more patrol officers onto the streets to work with the our citizens in their neighborhoods.

Chief Ramsey has explained more fully the times, locations, deployment of these cameras, but I want to reemphasize that the cameras are an extraordinary tool for extraordinary times. We are not monitoring the streets of the District around the clock and listening in on people's conversations. We don't even have the capability to do that. We are not tracking the movements of individuals around town or in private buildings. We don't have the capability to do that. Nor are we using video to identify individuals such as through biometric imaging. We don't have the capability to do that, either. Rather, this is a prudent, limited and legal use of video technology in support of our goal of ensuring peace and public safety during extraordinary times. Mayor Williams believes that we owe it to our residents, workers and visitors to be vigilant, innovative and careful in how we pursue that goal. Therefore, we monitor a limited number of public spaces during special events or at times of heightened alert as announced by the Federal Government.

As Chief Ramsey has explained, MPD is currently finalizing its operating policies and procedures for those uses for the Mayor's review.

I thank you for the opportunity to testify. I would be happy to answer your questions. And again I apologize for being late.

[The prepared statement of Ms. Kellems follows:]

HEARING ON
PRIVACY vs. SECURITY: ELECTRONIC SURVEILLANCE IN THE NATION'S CAPITAL

SUBCOMMITTEE ON THE DISTRICT OF COLUMBIA
COMMITTEE ON GOVERNMENT REFORM

CONGRESS OF THE UNITED STATES

CONSTANCE A. MORELLA, CHAIRWOMAN

TESTIMONY OF

MARGRET NEDELKOFF KELLEMS, ESQ.
DEPUTY MAYOR FOR PUBLIC SAFETY AND JUSTICE

MARCH 22, 2002

9:00 A.M.

Good Morning Chairwoman Morella, Congresswoman Norton, members of the Subcommittee, and all the distinguished guests here today. I am Margret Nedelkoff Kellems, Deputy Mayor for Public Safety and Justice for the District of Columbia.

Thank you for convening this hearing on the high-profile and often misunderstood topic of the District's use of video technology as a tool in ensuring the public's safety. I would like to take just a few minutes to outline the Mayor's priorities and concerns regarding the use of this powerful tool, and what steps we are taking to ensure that our practices are judicious and relative to our needs. Chief Ramsey will discuss the specifics of the current operation of the system and the policies and procedures that MPD is drafting to govern its operation.

Mayor Williams has consistently identified neighborhood safety and quality of life as among his top priorities for his Administration. Since the events of September 11, homeland – or in this case hometown – security has taken on a new significance, particularly in our nation's capital. But Mayor Williams continues to stress that in our pursuit of these goals, we must take great pains not to unduly impact the personal privacy interests of our citizens, the many people who work in our city, and the more than 20 million people who visit annually.

Video technology is not a new law enforcement tool. In fact, law enforcement has been using video technology for years as a means of collecting criminal evidence. As you may well imagine, without video surveillance, MPD would not have had nearly the success it has in closing down many of the drug markets that impact our city. Of course, what we are talking about today is substantially different than using video in targeted law enforcement efforts, and we appreciate the opportunity to discuss this issue in an open forum with you today.

As video technology has moved to the forefront as an effective tool in ensuring the public's safety and protecting against threats to our city, Mayor Williams has committed to an open dialogue and discussion of the benefits and concerns on both sides of the issue. We recognize that engaging all of the stakeholders in the development of our system – both the security experts and those seeking to protect our privacy rights – is the only way we will come to an acceptable result that everyone can understand and live with.

The Mayor has been very clear, and all of the stakeholders agree that the primary objective of the system is to enhance public safety during major events, times of heightened alert, and actual emergencies whether they are terrorism-related or not. During these times, law enforcement resources are our most valuable commodity in terms of ensuring safety and peace. Video technology is tremendously useful in helping us allocate and manage those resources effectively. Instead of relying on radio call-in information from officers scattered around the event or city – officers whose field of vision is limited to their immediate surrounding – video technology allows us to monitor large and distant areas quickly and unobtrusively. That information allows us to redirect officers, know where reinforcement is needed, and anticipate where we might need other types of equipment or response resources. This information assists in protecting the public, as well as our first responders.

In fact, the utility of video technology at major events was proven before September 11th. MPD leased video technology equipment during the IMF / World Bank demonstrations, the Inaugural, and even the NBA All-Star Game to assist in resource deployment during the events. Based on those successes and in anticipation of the planned IMF / World Bank meetings and demonstrations scheduled for the end of September last year, MPD began the development of a small video network capability in the Joint Operations Command Center. Since September 11th, MPD has expanded the video network to approximately one dozen cameras, focused on key areas of potential terrorist threats, and has been pursuing linkages with video systems operated by area transportation departments and Metro.

Even while we pursue the use of video technology in those kinds of situations, Mayor Williams has asked MPD to explore the possible future uses of this technology for controlling crime on a daily basis. Of course, that is a big step from where we are now, and there are many issues that must be considered and evaluated before we move forward. Questions regarding individual privacy rights as well as important operational concerns like the location of the cameras, how are they monitored – by whom and from where, are the video feeds tape recorded for evidence, and how long and for what purposes are the tapes maintained, would have to be worked out in great detail and MPD is already engaged in that process.

It is important to note that the video surveillance system is not intended to be a primary neighborhood-policing tool. There is no substitute for the security offered by the presence of officers patrolling neighborhoods themselves. That said, we must be responsive to public concerns about the potential uses of video surveillance once that capacity is in place. Several neighborhoods have requested that cameras be installed in their communities, hoping that it would have a deterrent effect on criminal activity, and perhaps even assist is capturing and prosecuting criminals. Certainly there will be disagreement among different communities regarding the appropriateness of video surveillance. Bear in mind that at the core of MPD's policing philosophy is a respect for the interests of communities. Chief Ramsey's Partnerships for Problem Solving model has as its core value that the police are to work with communities to solve problems together in a way that is acceptable to all involved.

But the fact remains that in some places – certain communities, stores, ATM machines, around government buildings, for example – video monitoring equipment has helped control crime and convict criminals. Mayor Williams has instructed us to assess the benefits and burdens of these applications to determine whether on balance, video technology as a community crime fighting tool is effective, even as we move increasingly more patrol officers onto the streets to work with our citizens in their neighborhoods.

Chief Ramsey will explain more fully the times, locations, and deployment of these cameras, but I want to reemphasize that these cameras are an extraordinary tool for extraordinary times. We are not monitoring the streets of the District around the clock, and listening in on people's conversations. We don't even have the capability of doing that. We are not tracking the movements of individuals around town or in private buildings. We don't have the capability to do that. Nor are we using video to identify individuals, such as through biometric imaging. We don't have the capability to do that either.

Rather, this is a prudent, limited, and legal use of video technology in support of our goal of ensuring peace and public safety during extraordinary times. Mayor Williams believes that we owe it to our residents, workers, and visitors to be vigilant, innovative, and careful in how we pursue that goal. Therefore, we monitor a limited number of public spaces during special events or times of heightened alert, as announced by the federal government. And as Chief Ramsey will explain more fully, MPD is currently finalizing its operating policies and procedures for those uses for the Mayor's review.

I thank you for the opportunity to testify before you today and I would be happy to answer any questions you may have.

Mrs. MORELLA. Thank you, Ms. Kellems. I am going to ask you, as I start the round of questioning for about 5 minutes and we will rotate, to answer as succinctly as you can so we can get as much information as we can in a short time.

First of all, to Councilwoman Patterson. The council has approved, I think it was emergency legislation requiring the submission of policies and procedures for the system for approval. I am curious about when you will be receiving those policies and procedures. Even more than that, what standards will the council use in evaluating the adequacy of the policy and procedures, especially for the public areas that involve Federal facilities in the Nation's Capital area.

And then, connected to that, does the council plan to consider any legislation adopting standards for the use of stored video information that would be gathered through electronic surveillance?

Ms. PATTERSON. Thank you, Congresswoman.

I believe that Chief Ramsey has indicated that the department is developing draft guidelines for use of the technology. It is those guidelines that the council, by emergency legislation, asked to come to the council for active review.

I am not sure what the timeframe is. I know that the chief had indicated a 30-day turnaround time in meeting with the ACLU during the tour of the facility that he mentioned. In terms of what we will then do, I anticipate that we will consider whether it is appropriate to do legislation, and I think it may well be.

I would just note that our own citizens are a bit ahead of us. As I noted in my testimony, the Federation of Civic Associations came up with a list of issues that they would like to see addressed in legislation, including civil and criminal penalties for abuse, and providing for destruction of tapes not being needed for evidence and such. I think these are issues we should look at.

As to standards, I am not familiar with the two sets of national standards that I understand have been developed. But I am delighted to know that they exist, and we will certainly be reviewing the ABA standards and the other standards that have already been developed on these issues to see what we may learn from those standards.

Mrs. MORELLA. I would like to ask, following up on that, Chief Ramsey, has the Metropolitan Police Department reviewed the standards that were recommended by the American Bar Association or the guidelines that are recommended by the Security Industry Association and the International Association of Chiefs of Police? For instance, both recommend that notice be given to subjects that they are under surveillance. That ties in with the committee input and what are we doing about letting the community know when they are going to be under surveillance.

Mr. RAMSEY. Yes, ma'am. In addition to that, we also have gotten guidelines that are being used in other countries to try to take a look at best practices as it relates to the use of CCTV as we draft our own policies. We do have a draft directive now. I have seen that. I sent it back for further work. And it is being staffed now by other members of the department, including our general counsel, and I hope to get that back within the next couple of days, and then we will be getting it over to the Mayor's Office and to the council.

So, yes, there is a lot that we are reviewing and looking at. Just yesterday the Mayor's Office, the deputy mayor, myself, the Mayor, the U.S. Attorney, spoke with officials from Sydney, Australia and talked a lot about their policies and their guidelines, and they have sent that to us to review as well. So we are really reaching out as far as we can to look at best practices.

Mrs. MORELLA. So you believe that you will have some kind of policies that you will be recommending and practices that you feel are appropriate that you will be submitting to the Mayor and the Council.

That will include how your community—how you advise the community.

Mr. RAMSEY. Yes, ma'am.

Mrs. MORELLA. It just seems to me that's a very important facet, that people are not taken by surprise and have justification.

Mr. RAMSEY. Yes, ma'am.

Mrs. MORELLA. Right now I think you said you have 12.

Mr. RAMSEY. Yes, ma'am. We have 12 cameras that are owned by us. They are primarily locations we feel we need to keep an eye on in the light of terrorist threat is primarily what we're concerned about right now. And we also had some problems early on, if you recall, about having officers actually assigned to certain locations. What this does is act as a force multiplier, it allows us to be able to keep an eye on certain areas that we feel are very vulnerable but not having actually to put an officer there.

Mrs. MORELLA. What Federal agencies have allowed you to install cameras on your facilities to survey public areas?

Mr. RAMSEY. Well, I'd have to get back to you with a complete list. I know one of our cameras is mounted on the Department of the Labor, the roof of the Department of Labor right next door to us at 300 Indiana Avenue. We can get a very good shot of the Capitol from there as well as the Washington Monument. Union Station is another location where we have a camera mounted. The Old Post Office, that's right, the Old Post Office. I do have a list here that I can provide to you.

Mrs. MORELLA. I think you've given us a list, Chief Ramsey. I'm curious about who in those agencies granted permission, and is there any written agreement among the agencies of how the cameras would be used and how long it will be stored, anything like that?

Mr. RAMSEY. I don't believe there's any written agreement, ma'am. And I would have to check with Steve Gaffigan who heads my Office of Quality Assurance to find out who specifically they spoke with there, but we did get permission prior to putting any cameras up.

Mrs. MORELLA. We would be interested in that, too, that gets into the policies and procedures for it. My time has expired. I'll get you on my next round, Mr. Parsons and Ms. Kellems. So now I defer to the ranking member for 5 minutes.

Ms. NORTON. Could I ask how these cameras first became known to the public?

Mr. RAMSEY. Actually, ma'am, they became known during the April 2000 IMF World Bank demonstrations when we first used it.

Ms. NORTON. No, no, no. That's not my question. I'm trying to get at something else. I want to know when they became known. For example, when Ms. Patterson first knew of these cameras. I mean, I would consider her an example of the public knowing as opposed to the police department.

Ms. PATTERSON. I appreciate that.

Ms. NORTON. How did you come to know about it?

Ms. PATTERSON. I think I could—I know the extent of the capacity from a tour that I did of the police command center fairly recently within the last 6 months, the center that became operational right after September 11th.

Ms. NORTON. I'm talking about the surveillance of citizens.

Ms. PATTERSON. I became aware of that in the course of concerns expressed by some of the demonstrators in 2000 and 2001. They expressed concerns about the issue. But in terms of knowing——

Ms. NORTON. Of course, that's during a demonstration. I'm talking about the use of cameras, the use of the network to surveil people who are not involved in demonstrations. When did the council—when was the council put on notice that this was happening?

Ms. PATTERSON. I'm not sure we have to date been put on notice that precisely what you described is happening. Because it's my understanding from the deputy mayor and the chief that we are not as a matter of course using the technology apart from major events.

Ms. NORTON. Well, that brings me to my next question. The chief testifies that we're using this in major events. We're going to use it in an emergency. Mr. Parsons testifies that the Park Police personnel will continuously monitor the cameras at a Park Police facility. These are, of course, cameras that go into our command center. Ms. Kellems command, as you may well imagine, without video surveillance MPD would not have had nearly the success it has had in closing down many of the drug markets that impact our city. Mayor Williams has indicated that this city should have a system such as the one they have in England with—and, of course, that involves 2.5 million cameras.

I'm real confused here about what appears to be the difference in Federal policy, city policy. I'm confused about the notion of notice. It's one thing to—it's one thing for demonstrators to find out they're being surveilled. It's another thing for the council to be given a tour. It's another thing to have notice and comment so that ordinary members of the public can know about what amounts to a major step that it seems to me people are at least entitled to know about before, not after the cameras are already set up in spots. First I see policy differences, and next I want to know why there was no notice and comment before this was done.

And finally, Ms. Kellems, you need to explain what the Mayor wants to do, since you represent him, whether we're on our way to London and a video on every corner. Please don't include 9th and East Capitol by where I live, by the way. Save yourself some money on my plot. But go ahead.

Ms. KELLEMS. I'll start and I'll let the chief speak to it. I, of course, can't speak to the Federal policies about how often they're going to be using these and monitoring these.

Ms. NORTON. But you haven't coordinated that with them. You heard his testimony today. And so what they do, they can do as much as they want to with your system.

Ms. KELLEMS. Sorry, it's their system and they're doing this independently. We have, on occasion when we've brought up our command center for major events, asked if we can access their feeds. We do not have this on a normal ongoing basis. We don't monitor their system at all. They monitor it separately.

Ms. NORTON. What is the relationship between their system and your system?

Ms. KELLEMS. They're separate.

Ms. NORTON. No relationship to the command center?

Ms. KELLEMS. We have the ability when we want to access their system to do that, which we have limited to public events and vice versa. We do not have our command center up and running and monitoring their feeds except during major events.

Ms. NORTON. Go ahead.

Ms. KELLEMS. As for the comment about the drug market, I can let the chief explain it more fully. I was speaking in generalities that I probably shouldn't have. Those are not mounted, permanent CCTV cameras. I was speaking to the use of video technology when the police are engaged in an activity around a particular drug market. That is a very different type of technology, and those feeds are not going into our command center. I was just referring to the use of video generally.

The Mayor's position is that to the extent these are useful tools, he would like MPD to give him some advice and do some research on how they're used other places, and London is among them. As the Chief mentioned, we were speaking with the folks from Sydney, Australia last night. Their policy is exactly the opposite of ours. They do not use them for major demonstrations, they use them in crime hot spots. They claim, and they are sending us the information, that has been an effective deterrent tool. But we're very much in the investigative point. We have not moved into using these on a regular ongoing basis or using these in the neighborhoods at this point.

Ms. NORTON. Mr. Parsons, on the basis of what evidence have you decided that 24/7 around-the-clock surveillance is effective?

Mr. PARSONS. We are convinced by studies and consultants that these icons of democracy are high targets for terrorist activities. And that is the sole reason that have made the decision to go forward with planning for these cameras.

Ms. NORTON. My time is up.

Ms. PATTERSON. Madam Chair, could I respond to that portion of the question that was directed to the council, with your leave?

Mrs. MORELLA. Not on my time, but yes, go ahead.

Ms. PATTERSON. Thank you very much. There is an issue, I think, between what is operational practice by law enforcement and what is public policy where the legislature, on behalf of the public, needs to be involved. And I think you might find a distinction of opinion even on this panel on where that line is drawn.

The council, I think it's clear, should not sign off on an individual video stakeout of a corner drug market, but the council should sign off on what is our policy for maintaining videotapes from demonstrations, for example. So I think there is a little tension here in terms of what belongs to the policymakers and what is day-to-day law enforcement operations.

Ms. NORTON. Of course, there were no policies in place whatsoever when this surveillance first began. And there is no written policy in place as we speak.

Ms. PATTERSON. Right.

Mrs. MORELLA. I think that is critically important that we bring it all together and find out how one links to the other, the nexus. Chief Ramsey, did you want to quickly comment on that one?

Mr. RAMSEY. I just think it's a very important point to note that to date we haven't been recording anything. And we haven't because we don't have the policies in place. So this is simply observation at this point in time. They're not being recorded. We have the capacity for recording, but we have not recorded. And I just think that's a very important point to make. Because we understand the sensitivity.

I don't disagree with council member Patterson that the role of the council is to help us and to pass legislation that helps set some of these guidelines so that we can safeguard against abuse. That's what we're trying to do now with draft policies, reviewing all the literature that's out there now around how different cities and, indeed, different countries handle these very, very sensitive issues. But we have not recorded anything nor will we until we get a policy in place.

Mrs. MORELLA. Did you record during the World Bank and the International Monetary Fund——

Mr. RAMSEY. No, ma'am. No, ma'am.

Mrs. MORELLA. Mr. Parsons, on that issue, what does it mean as you stated in your testimony that images will be recorded on a continuous loop?

Mr. PARSONS. Continuous loop means a device that can be recorded over itself after a period of time. Continuous loop tape.

Mrs. MORELLA. So you are recording.

Mr. PARSONS. Well, at the current time we have no cameras installed.

Mrs. MORELLA. You are planning to.

Mr. PARSONS. We're still in the planning stage. That is our intent, yes.

Mrs. MORELLA. What standard do you plan to use to determine when the recordings will be erased? I know you told me you haven't done it yet, but when you do, what are you going do to in terms of how long you're going to keep them and when they will be erased?

Mr. PARSONS. This is all part of the planning process. We haven't really come to the point of determining that.

Mrs. MORELLA. I think it's probably a good thing that we have this hearing now, too, because we can kind of prompt you to see why you need to work together in partnership to see what is—I mean, for instance, as we get back to the Metropolitan Police Department, is the Federal Government at the monumental core going to link up with the command post, the command center in the District of Columbia? Is that something you're thinking about?

Mr. PARSONS. That would be our intent, yes.

Mrs. MORELLA. That would be the intent, that it would be connected. Let me ask you, also, since you've talked to Australia and London and other countries, how effective are cameras in combating crime or preventing terrorism? What were the results of the research that you may have read about or undertaken? Do you believe that cameras are a deterrent to crime? And if that, in fact, is the case, can we expect that there will be a reduction? We've been reading a lot about crime in the newspapers recently. Will there be a reduction in crime where they are deployed?

Mr. RAMSEY. Well, if I may, certainly cameras are not a panacea. They're not going to end all of our problems. But one thing that we learned yesterday in our conversation with officials from Sydney, 10 percent of their arrests are the result of their CCTV system, 10 percent of all their arrests. That's a lot. And last year, if my memory serves me correct, and I was taking notes yesterday, so hopefully I took good notes, but 735 assaults, arrests for assault were made last year as a result of these cameras. Again, that is, I think, something that we have to pay attention to. Again, these are street assaults that are taking place. Again, as far as we're concerned—I'm sorry, that was 783 assaults since January 2000. So that actually covers a 2-year span. I just found my notes from last night. But they did say that 10 percent of the arrests that they make come from that system.

So, again, does it stop all crime? No. Do they have it posted? Absolutely, right in the neighborhood, clearly posted. This neighborhood uses video CCTV. And we would be looking to do the same here, is have it posted.

Mrs. MORELLA. You know, jumping into another subject that you reminded me of, there are no postings right now in the District of Columbia for any of your traffic surveillance cameras. I've heard from constituents and District of Columbia residents who said I had no idea that this was an area with surveillance cameras.

Mr. RAMSEY. Ma'am, if I may, the red light cameras are posted. The photo radar right on our Web site, we put the locations of where we're going to be working. There is no way to put a sign up because those are mobile units. But we do post it on our Web site.

Mrs. MORELLA. For exceeding the speed limit, some of those?

Mr. RAMSEY. We actually post where we're going to be.

Mrs. MORELLA. I guess people aren't looking. I'm going to start looking myself now.

Mr. RAMSEY. They aren't looking at the Web site, and in some cases they're not looking at their speedometer, either.

Mrs. MORELLA. I have people who have contacted us and said, you know, I got this $100 ticket and I didn't even know that I was exceeding the speed limit. How do I appeal? Do I appeal a camera that records there?

The District's Department of Transportation plans to install 100 traffic cameras around the District. I would like to direct this question to Deputy Mayor Kellems. Does the District plan to link those cameras to the system?

Ms. KELLEMS. We'd like to. Those cameras are intended for traffic management particularly during an emergency. But we'd like to use them at all times to just sort of manage our terrible traffic situation here. We're planning to put them at major critical intersections and some of the commuter routes in and out of the city.

We think it would be useful to have that if we had, for example, an incident like we had on September 11th when we're trying to evacuate most of the people from downtown or they're self-evacuating, that would be useful information. It's police officers and National Guard who are usually deployed to the intersections to manage the traffic.

Mrs. MORELLA. My time is expired, but, Mr. Parsons, did I hear something about 700 new cameras that would be employed in the District of Columbia that would include areas where the Park Police has jurisdiction?

Mr. PARSONS. We are.

Mrs. MORELLA. Like Rock Creek.

Mr. PARSONS. We have been approached by the Department of Public Works with a map of the city that shows, as we understand it, 700 cameras. Very few of those are on Park Service jurisdiction. They would monitor traffic up along Rock Creek to the zoo, for instance, Rock Creek and Potomac Parkway. We're just in the preliminary stages but my understanding was 700 cameras.

Mrs. MORELLA. Who gives authority? I don't know from whence the authority for establishing these cameras comes, different authorities? And do they ever come together or is that something we're going to be looking at in the near future?

Mr. RAMSEY. Ma'am, as far as law enforcement goes, we do come together on a regular basis. In fact, this afternoon through the Council of Governments I'm participating in a meeting of all the regional law enforcement agencies around our regional emergency response plan. And we talk about a variety of issues including being able to work together through our joint command center. So there is ongoing dialog.

And I think it's important to maybe clear up one issue as well. Our goal is to tap into existing CCTV networks. But in order to do that, we just can't just start surfing the area and looking for images. We have to have the cooperation of the other agencies. The schools are a good example. If they don't turn on their switch on their side, we can't get that feed. And we would only do it in the event of an emergency inside a school.

Mrs. MORELLA. So you can tell them to do it, and they should report to you.

Mr. RAMSEY. Right. We can ask them and request and they would more than likely be making the request of us, quite frankly. Those areas during a time of a terrorist threat, our agencies would get together and we would provide access to one another automatically because of the nature of the threat so they could get our feeds, we could get their feeds. But we're in a state of heightened alert, and it would make sense to do that.

Mrs. MORELLA. Would you use a criminal justice coordinating council?

Mr. RAMSEY. That is certainly another vehicle that could be used. It hasn't yet, but it certainly could be used in that fashion. Although many of the agencies that are represented in this CJCC aren't necessarily those agencies that would have camera networks.

Mrs. MORELLA. You know, it would be awfully helpful to this subcommittee if you could in some way give us some kind of a chart to show how this all works out. Maybe you could all work together in doing that.

Mr. RAMSEY. Yes, ma'am.

Mrs. MORELLA. I now recognize Ms. Norton.

Ms. NORTON. Thank you very much. The Chair's reference to the CJCC is a very important one because even though not all the agencies are included, essentially what you will have in this city is the beginning of a model for what is going to happen elsewhere, and it may become a model for how people coordinate.

Let me make my position clear. I think you are on the horns of a dilemma. And I think you are trying your best to sort out this situation. In the process, it seems to me, the difficulty I have with what the Mayor said about London and the rest of it is that he does not show the appropriate balance. It does seem to me that every step of the way the second question has to be said, how do I limit this, how do I control this. And that is a concern I have about the testimony here. I don't hear that part of the equation being raised except through the council hearings.

I want to ask who has paid thus far for the D.C. cameras and for the link of the D.C. cameras to the Federal facilities, the White House and so forth, who has paid for that and out of what funds, out of what part of the budget did it come.

Mr. RAMSEY. The funding came from our capital IT as well as our capital, I guess, facilities funds, you would call it, has funded this thus far, not from our local budget.

Ms. NORTON. Do you expect that the homeland—that—we got a rather handsome homeland security amount. Do you expect that this future funding will be coming out of homeland security funding?

Ms. KELLEMS. I think that would be an appropriate use of it. In fact, the cameras we were referring to from the Department of Transportation were funded through the Federal appropriation for emergency preparedness.

Ms. NORTON. Mr. Parsons indicated that essentially what is being done here is to surveil public spaces, not individuals. Mr. Parsons, since you are surveilling public spaces and not individuals—that is somewhat comforting—what exactly are you looking for?

Mr. PARSONS. These cameras we are planning for would be inside the chamber of the Lincoln Memorial, for instance, and also focused on the entry stairways.

Ms. NORTON. This is important. You're inside.

Mr. PARSONS. Correct.

Ms. NORTON. All right.

Mr. PARSONS. And focusing on the approach stairways. It will simply be there to anticipate anything that could occur on a 24-hour basis.

Ms. NORTON. Mr. Parsons, is the Park Service also working on rules and guidelines that would help your personnel know how, where, when, etc?

Mr. PARSONS. Certainly as it relates to how long we're going to keep this information, yes.

Ms. NORTON. And how long are you going to keep this information?

Mr. PARSONS. We're not there yet. But certainly this hearing has brought us to the realization that we need standards and policies.

Ms. NORTON. You're absolutely right about how long you're going to keep the information, where they should be—the reason I ask, Mr. Parsons, is you indicated that apparently different Federal facilities have different policies. You went down a list of park department facilities, some of which—some of whom were using, some of whom were not. So there is obviously not any uniformity within the Federal Government. Do you intend that all Park Service facilities will have the same view of these——

Mr. PARSONS. That's our objective, yes.

Ms. NORTON. Do you anticipate that eventually all Park Service monuments will have these cameras throughout the United States?

Mr. PARSONS. I don't believe so. Only those that have been identified as a high threat.

Ms. NORTON. Could I ask about the schools? Excuse me, Chief Ramsey had——

Mr. RAMSEY. Ma'am, I just wanted to add one thing from MPD's perspective. Right now in our draft policy we're looking at 72 hours as a period of time if anything was recorded that would be kept unless there was some criminal activity, which would then—we would obviously maintain it longer for evidence.

We are looking at—and yesterday we found that in Australia, for example, they keep theirs for 21 days. Why? Because they want a period of time if there is an audit or something there for people to review the tapes. We fully support an audit process, and that's going to be part of our directive and something we work with the council on to see to it that it takes place in our system as well. I believe that if someone calls and says, hey, we want to see the tapes from a given day, we just hold those tapes. So I don't know if you need to really hold them 21 days or not, but I think we need to have a timeframe that is sufficient for review by interested parties that want to review it and not keep them too long. So that is where we're at as far as our policy goes.

As far as what we'd be looking at if a truck pulled in front of Union Station in an area where trucks aren't supposed to be and is left unattended, we can get on the phone and call security there and say, hey, you better go check out that truck and dispatch a scout car because it could possibly be a truck bomb. That's the kind of thing we're looking for, unusual or suspicious activity that would necessitate the dispatching of a scout car to check it out further.

Ms. NORTON. Almost inevitably then the police department is going to be sharing this information with Federal authorities, almost inevitably.

Mr. RAMSEY. Absolutely, ma'am, because the Union Station example, we wouldn't dispatch a scout car unless there was some reason why we couldn't reach the security that is already at Union Station. And that's part of your plan with the legislation you moved around the cooperative effort between agencies. They're part of that agreement. They would respond to that request.

Ms. NORTON. My time has passed.

Mrs. MORELLA. We both seem to be on the same wave length, Congresswoman Norton and myself. Picking on up on what you said, Chief Ramsey, who would be the interested parties?

Mr. RAMSEY. That is something that, ma'am, it could be the council, it could be ACLU, it could be you, it could be someone who wants to take on—we would have to work that out through some kind of process where we would agree that some group could come in and spot-check and take a look and see what we're doing. We're not trying to hide anything here.

Mrs. MORELLA. And, actually, I really meant it just the opposite. Is there going to be some criteria you will establish to gain access? You wouldn't want anyone who was just curious to have access to these records.

Mr. RAMSEY. Absolutely, it would have to be.

Mrs. MORELLA. You said you keep these records or the camera tapes for 72 hours.

Mr. RAMSEY. That's the draft policy right now. We haven't been recording. But the time span we're looking at right now or at first blush was 72 hours. That's not saying that there was any—anything about 72 hours that was—I mean, I've seen some policies that 96 hours—yesterday we learned 21 days. We need to find something that's reasonable here, work with the council, if they find it reasonable and all parties find it reasonable, and then put in place a mechanism where if there were someone who were the auditor, let's say, to come in and periodically review these tapes, there would be sufficient time for them to be able to do it and we would simply hold it.

Mrs. MORELLA. And to Mr. Parsons and the Park Service, I have a feeling that you're going to say this is work in progress, because I would ask what standards will you use to ensure privacy rights are protected when you install the cameras at various facilities?

Mr. PARSONS. I'm afraid I'm going to have to give you the same answer.

Mrs. MORELLA. So it's good that I pose it for the record, too, that you can look into the privacy concerns. But I will also pick up on we're going to have in our next panel imminently the American Bar Association, and we have already talked a little bit about the fact that they have come up with some standards. I wonder if you're familiar with them. Do you plan to use them when you're installing the security devices at the monumental core and the park sites?

Mr. PARSONS. No, we have not studied them at all.

Mrs. MORELLA. OK. Great. Then you really have an agenda before you. I hope that you will report back, I hope all of you will report back to this committee, knowing that what we're trying to do is achieve a balance. This was really kind of a fact-finding hearing because we think it's exceedingly important to know where, why, how long, standards, privacy, community input. And so, Ms. Norton, that was really my final question.

Ms. NORTON. Just a few more questions, Madam Chair. Chief Ramsey, I have staunchly defended your red lights here in the Congress and another one of my committees. I was quick to indicate that you have saved lives in some substantial numbers, that people in the District have clamored for them because people were being hit by cars and otherwise injured, and because you had limited the system so that, of course, you took only a picture of the license plate. There were some valid criticisms, the council raised those, the criticisms of who was running it, how people got paid. You're attending to that. That's the kind of model of what we'd like to see here. Nobody expects us to get it all right in the first time. No society has used these kinds of cameras before. The real question is limits and protections.

I'd like to ask about the use of the cameras now. I was comforted that you used them during the Winter Olympic games, you used them during times when there were clear—there was a clear need. There has been an attack. And I'm not willing to assume there will never be another attack. I wonder if you are looking to use the cameras when they are specifically—in a specific emergency, whether you are coordinated now with the Federal Government's new color-coded alert system, so that we all know when an alert—what an alert means?

Mr. RAMSEY. There was, I believe it was a 60-day period for comment with this particular system that Governor Ridge laid out for us just a week or so ago. We are looking at that. I don't have any, you know, cutting edge comments around it, quite frankly. I think the more important issue is that we all be on the same page.

Ms. NORTON. That's really my only question. When they come to an understanding of the notice and comment page—by the way, they did it the right way. They did notice and comment so we all know about this so you can have a say. At that time you will calibrate yours to the Federal.

Mr. RAMSEY. Yes, ma'am.

Ms. NORTON. Let me ask you, Chief, about an example you gave that is intentioned with the notion of emergency use only. You talked about an emergency at Union Station and you would be able to send somebody there. Well, that indicates consistent use of the cameras of the kind that the Park Service is doing as opposed to use of the cameras only in emergencies. So are the cameras, for example, at Union Station, which I take it are our cameras, are they going to be on the same 24/7 basis that the Park Service cameras are?

Mr. RAMSEY. No, ma'am. I was referring—and I apologize for the lack of clarity. I was speaking in terms of when the cameras are activated during this period of time, should we observe this kind of activity, then we would do that. Otherwise the cameras would not be on. Today, for example, the joint command center is being used for training for our crime mapping which has no camera capability at all, but it's our new crime mapping system. We have crime analysts in there learning how to use the new system. So it's really a multipurpose room, not just for CCTV.

Ms. NORTON. I must ask about the schools. First of all, if these are going to be hooked up to schools, does that mean we're going to have a camera in every classroom, in every auditorium and in every hallway? How could we possibly hook these up to schools?

First let me ask Ms. Patterson, how were they used at your daughter's school?

Ms. PATTERSON. The closed circuit TV at Alice Deal Junior High School that was put in place in 1996 was a security measure so that you didn't have to have people walking all over the school grounds at any particular time just as you do at any kind of closed circuit television security system in a public building, to enable one person to sit at a bank of cameras and see the full exterior of the school. Since that time the school system has moved to that kind of technology, I believe, through the junior and senior high schools first and then moving into the elementary schools.

I'm not familiar with what is planned for cameras internal to the buildings of the D.C. public schools, although I know they do at least have one camera inside the office of the school system. So I'm not quite sure what they're intending to do on the interior of buildings.

Ms. NORTON. Perhaps Chief Ramsey or Ms. Kellems could tell us what your intention is with respect to schools in particular.

Mr. RAMSEY. If we were to have a Columbine-like incident, we would request the ability to be able to tap into their CCTV system.

Ms. NORTON. Their existing system?

Mr. RAMSEY. Their existing system that they have in the school. And that would help us know where the gunman was located in the school, for example. Where do we need to get kids out, what are the evacuation routes that would be best taken in light of where the incident is occurring versus where students could be safely evacuated. What routes should our cars take approaching the scene? If the gunman is in the northeast corner of the building, we don't want our cars coming in from the northeast because they— the sniper fire. So it would help us to be able to isolate the incident a lot easier than having our emergency response team go in and having to do a physical search to get the information an hour and a half later that they could have gotten in moments.

Ms. NORTON. This is an important distinction. If these cameras exist, we're talking about halls, we're talking about the kind of cameras that are already in use. We're not talking about an add-on bunch of cameras in schools.

One of the dilemmas we face here, I'll say finally, is that, to take the Union Station example, if the chief had the manpower to put 100 cops there to look, then nobody would say that there was anything wrong with that kind of surveillance. If what we have instead is a camera that replaces the 100 cops, the dilemma we face is, is that any different. We know that because it is surveillance there are some differences.

One of the standards we might use is to try to equate as far as possible the kind of surveillance that police ordinarily do with the kind of surveillance we're talking about here. If anything, the kind of surveillance that my neighborhood policeman does is probably more invasive. He probably looks at my face, he probably remembers things about me. But we are used to him. This is considered a part of ordinary law enforcement. What we're going to have to come to grips with is what difference it makes when we make this cosmic leap and whether we need special safeguards in light of that kind of leap.

I thank you very much, Madam Chair.

Mrs. MORELLA. Thank you, Congresswoman Norton. I want to thank you for being here. I do want to ask you to report back to us with regard to the standards that you are establishing, even as you consider them, your time schedule, policies, and public input opportunities and plans. And we would appreciate that. We'll also get to you some questions about the red light cameras which we didn't feel was the focus of this hearing today.

But I do want to thank all of you for being with us, Ms. Patterson, Ms. Kellems, Chief Ramsey, Mr. Parsons, and I'm going to ask that there be a 5-minute recess and then we'll get to our second panel. Thank you very much.

[Recess.]

Mrs. MORELLA. I'm going to call back into hearing our second panel as we resume the subcommittee hearing. Thank you again for your patience, too, and for being here. Johnny Barnes, executive director of the American Civil Liberties Union of the National Capital area, Ronald Goldstock of the American Bar Association, and John Woodward, Jr., of RAND. So gentlemen, may I ask you to stand and raise your right hand also to be sworn in.

[Witnesses sworn.]

Mrs. MORELLA. Thank you. Again, affirmative responses by all of the panelists. So, again, if you would proceed about 5 minutes each with your testimony. And I think you probably noted that we alluded to some of the points that you're going to be bringing out with the previous panel, which demonstrates that your testimony was very, very helpful. So we'll start off then with you, Mr. Barnes. Thank you.

STATEMENTS OF JOHNNY BARNES, EXECUTIVE DIRECTOR, AMERICAN CIVIL LIBERTIES UNION, NATIONAL CAPITAL AREA; RONALD GOLDSTOCK, AMERICAN BAR ASSOCIATION; AND JOHN WOODWARD, JR., RAND

Mr. BARNES. Thank you, Madam Chair. I'm delighted to be here in these familiar surroundings, Congresswoman Norton. I gave roughly a quarter of a century of my life as a staff person to Congress sitting behind those seats in which you all sit very often. I can assure you that the perspective from this seat is very different.

You have my testimony I'd like to include in the record in its entirety, and I'll summarize.

Mrs. MORELLA. Without objection, it will be.

Mr. BARNES. The ACLU believes the District should abandon its plans to establish a British-style system of surveillance cameras. There are five compelling reasons that drive our belief.

Surveillance cameras are not effective at fighting crime. They don't work well. Surveillance cameras reduce resources for placing police into neighborhoods where they are needed. The money can be better spent. Surveillance cameras undermine individual privacy and are inimical to our way of life. We have Times Square in America, not Tienemen Square.

What are we doing to our icons of democracy? Surveillance cameras should not be contemplated without the permission of those they impact. Such permission has not been granted by the people of D.C. nor by their elected representatives, the D.C. Council.

Surveillance cameras are subject to great abuse. Young and minority citizens already face the burden of profiling by the police. They should not have to carry the additional weight of being singled out, tracked and traced by the camera's eye, simply because of their age or the color of their skin, and for no other reason. And women should not face the additional indignity of leering eyes peering through the lens of a camera. Yet that is the sad legacy that is being left where these cameras have been tried in other places around the world and across the United States.

I'll briefly expand on these five reasons.

Cameras do not make us safer. Instead they give the public a false sense of security. As you noted, Madam Chair, despite 2.5 million cameras in England, including the 150,000 cameras in London, where the average citizen is filmed 300 times a day, the murder rate in that capital city is at record levels, and street robbery, the very crime these cameras are supposed to prevent, will soar to 50,000 in this year alone.

Madam Chair, cameras don't catch crooks. Cops do. That is why Detroit, after a decade and a half, and other cities, many other cities throughout the United States, have abandoned this adventure of surveillance cameras. Surveillance cameras require an upfront investment in technology and require ongoing maintenance. In addition to monitoring the video screens, police officers must be pulled off the street and put into the video control room. A camera can't stop a mugging or a murder in progress on the streets. A police officer can. The use of video surveillance cameras goes far beyond a change in the style of life as we know it. The use of these cameras will change the Constitution as we know it. The terrorists are winning.

No longer will we feel free to sunbathe in our backyards because helicopters mounted with cameras can fly over and film residential neighborhoods. They are already equipped to do that. I believe, Madam Chair, those are additional cameras beyond the dozen or so to which Chief Ramsey referred.

No longer will we be able to have the expectation of privacy in a restaurant in Georgetown. Merchants there have asked to join this network of cameras. Let me just say, Madam Chair, I believe already you're making a difference. You're having an impact with this hearing. Because the representations, unsworn, not under oath, that have been made in recent weeks and months have been inconsistent with some made today. And so, as you point out, this hearing and hearings that will follow is effective in flushing out the facts and getting to the truth. Perhaps because we've had different spokespersons, the inconsistency is there. But it is there.

So powerful are these cameras that they can film a belt buckle a mile away. So strong is surveillance technology that some allow the viewers to see through clothing. We first learned of these cameras when we read in The Wall Street Journal on February 13th that the plan for their use, "go far beyond what is in use in other American cities." That's Mr. Gaffigan, the gentleman who's coordinating the program with the metropolitan police department. On that same day, Access Communications, the Australian firm that constructed the District's new system, told us that this system would be used, "in support of every-day policing." That's what the firm that constructed the system said in a press release.

General sweeping surveillance of all of us, rather than particular specific surveillance of those among us who act on terror, who commit crime. Mayor Williams confirmed that in an interview on March 8th in The Washington Times, and Chief Ramsey, the following day, March 9th repeated the assertion that the District would have a British-style system.

This hearing is very important, and the hearings that will follow are very important. We need answers. Is this system designed to help prevent terrorism or catch terrorists, or is this system to be used for general law enforcement? When will the system be on? Who will monitor it? Will the system use biometrics? Where, how long, and by whom will recording be kept? Will there be laws and regulations governing this system? We need a body of laws, a set of regulations, not policy guideline.

Today we do not know all of the answers to those questions. Without this and other hearings to explore these issues, tomorrow may be too late.

Finally, the abuse. Law enforcements have stressed this. Perhaps if Chief Ramsey or Assistant Chief Gainer monitored the system on a daily basis, we would more likely trust them. But the muddy history of this Nation has given rise to far too many occasions where the courts and our legislatures have enforced and imposed controls because law enforcement could not be trusted.

The young black attorney who knew his rights was nonetheless stopped and searched on a major highway because the police had a written memo, a policy guideline profiling, allowing profiling of young black men as probable drug dealers when they're in rental cars driving along I–95. D.C. police lieutenant followed and sought to blackmail gay men who frequented certain establishments, and law enforcement says trust us. Imagine the damage to lives and in other situations this could have caused if bad apples in law enforcement had the additional power of the camera's eye. These cameras will watch all of us, not just those bent on crime.

I close with a bit of history and perhaps an admonition that I believe presents interesting comparisons, if not striking similarities. In 1790, a Congress seeking security from a ban of Pennsylvania militiamen, sacrificed liberty and enacted article 1, section 8, clause 17 of the Constitution, the District clause. More than 200 years later, because of that provision, we still have a British-style system in the Nation's Capitol. I know you're sympathetic to that cause, Madam Chair. Thank you.

[The prepared statement of Mr. Barnes follows:]

STATEMENT

OF

JOHNNY BARNES

EXECUTIVE DIRECTOR

AMERICAN CIVIL LIBERTIES UNION OF THE
NATIONAL CAPITAL AREA

ON

SURVEILLANCE CAMERAS IN THE DISTRICT OF COLUMBIA

BEFORE THE

SUBCOMMITTEE ON THE DISTRICT OF COLUMBIA OF THE

HOUSE OF REPRESENTATIVES COMMITTEE ON GOVERNMENT REFORM

MARCH 22, 2002

My name is Johnny Barnes. I am the Executive Director of the American Civil Liberties Union of the National Capital Area. Our affiliate has approximately 6,000 members from Washington, D.C. and Montgomery and Prince George's counties in Maryland. The ACLU is a nationwide, non-partisan organization with nearly 300,000 members dedicated to protecting the individual liberties and freedoms guaranteed by the Constitution and the laws of the United States.

Thank you for the opportunity to testify today about the use of general surveillance cameras in the District of Columbia. In the wake of the horror of September 11th, all of us are extremely concerned about protecting public safety. Unfortunately, many of the proposals to enhance security have been grossly oversold. Surveillance cameras are one such bad idea. They would not make us safer and would undermine the individual liberties that are fundamental to what we are as a nation. The ACLU believes that we can be safe and free, that we do not have to destroy liberty in the name of security. If terrorists are successful in destroying our freedoms, terrorism will have won.

Surveillance Cameras in the District of Columbia

First, we must be clear about what we mean by surveillance cameras as contemplated by the District's Mayor and his Chief of Police. We are not talking about cameras that detect cars running red lights or speeding. While those systems are not without their problems, their cameras focus on specific offenses. They are only supposed to record individual violations. Red light and speeding cameras are not supposed to record the movements of law-abiding persons, those who respect traffic lights and who do not speed.

The surveillance system planned for Washington, D.C. is totally different. It would provide an extensive network of cameras throughout the city that would be in operation at all times. It would be indiscriminate in recording the innocent and the lawbreaker alike. The purpose of the system would be to enhance law enforcement in general, searching for all crimes, from so-called quality of life crimes to drug dealing to crimes of terror.

Any confusion about what the surveillance system for the District would entail was dispelled by the Mayor and the Chief of Police in recent interviews with the *Washington Times*.[1] The paper reported that according to Mayor Anthony Williams, "increased government surveillance is a reality of the post-September 11 world"; "the District needs to follow the lead of cities such [as] London and Sydney, Australia and expand its camera system." The Mayor was quoted as saying, "We are in a new . . . really dangerous world now, and we have to maintain a higher level of security."

[1] *Washington Times* of March 8 and 9, 2002, respectively.

Chief of Police Charles Ramsey echoed the Mayor's words. He said that the District "must and will expand its use of surveillance cameras, much like London, which uses 150,000 cameras to monitor its population."

At the outset, the critical point is that Washington, D.C. by adopting the British model, is not contemplating a system narrowly limited to fight terrorism. The British system was originally justified by the fight against I.R.A. terrorism. However, today there are about 2.5 million surveillance cameras across England used for general law enforcement. Thus by adopting the British model, the Mayor has effectively acknowledged that September 11th is being used as a license to establish a surveillance system that goes far beyond the fight against terrorism.

There are five principal reasons why the Mayor's proposal to set up a British-style surveillance system is a bad idea. The ACLU believes that these reasons argue conclusively that Washington, D.C. should abandon its plans to create such a system.

Reason No. 1: Surveillance Camera Systems Are Not Effective Crime-Fighters

Britain has the largest closed circuit television surveillance system in the world. Cameras are everywhere: on the streets, on buses and trains, in buildings, etc. The average London dweller can, in the course of a day, be filmed up to 300 times.[2]

British surveillance cameras, however, have not stopped crime. Examining the British experience, UPI reported on March 8, 2002 that "crime is soaring across the country. In London, a city of 8 million people, murder is going on at a record pace. Street robbery, the very crime that CCTV [closed circuit TV] is supposed to be best at deterring, will reach 50,000 this year." UPI noted: "A three-year study commissioned by the British government and conducted by the Scottish Center for Criminology suggested that 'spy' cameras had little or no effect on crime. It concluded that 'reductions were noted in certain categories, but there was no evidence to suggest that the cameras had reduced crime overall.'"

The evidence from Australia is equally unpersuasive that surveillance cameras are effective against crime.[3] At best, the camera system in Sydney that Mayor Williams wishes to emulate produced one arrest only every 160 days.[4] Before limited resources are spent on surveillance cameras, close attention must be paid to the claimed benefits.

---

[2] *The Guardian (London)*, *Guardian Society*, page 2, August 1, 2001.
[3] According to the Deputy Director of Sydney University's Institute of Criminology, *Sydney Morning Herald*, December 26, 2001.
[4] *Ibid.*

Closer to home, consider the experience of Oakland, California. For three years, the police department advocated the use of surveillance cameras in public places. The department had technology that could read the fine print on a flyer from hundreds of yards away, and that could recognize a license plate or a face from more than a mile away. In a report to the City Council, Chief of Police Joseph Samuels, Jr., stated that his department had hoped to be ". . . among the pioneers in the field of taped video camera surveillance" but ultimately found that ". . . there is no conclusive way to establish that the presence of video surveillance cameras resulted in the prevention or reduction of crime."

Detroit, Michigan took 14 years before it decided to abandon its surveillance camera system in 1994 citing high maintenance and personnel costs and mixed results.

The problem with taking a cold look at the claims made on behalf of surveillance cameras is that intuitively many people *feel* that cameras should be effective fighting crime. But "feel-good" is not the same thing as "do-good." There are many reasons why cameras do not work. Here are two: First, criminals learn to stay out of camera view. And second, when criminals conclude that they cannot do that, they go elsewhere, the so-called displacement effect, so that crime rates are not reduced.

Reason No. 2: Surveillance Cameras Displace More Effective Public Safety Measures

As feel-good measures, surveillance cameras lead us to waste limited resources that could be better spent putting police officers into neighborhoods. Surveillance cameras require an up front investment in technology and require ongoing maintenance. In addition, to monitor the video screens, police officers must be pulled off the street and put into the video control room.

Surveillance cameras are not a substitute for community policing. On March 14th, the *Washington Post* reported the "spate of violence" in the District. Six persons died and three suffered from gunshot wounds in five incidents over four days. D.C. Councilmember Jim Graham said that the Georgia Avenue shooting "underscores the fact that we need more officers in this neighborhood." Councilmember Kevin Chavous said that his constituents did not see the additional officers promised by police Chief Ramsey two months ago. An extensive camera surveillance system would require officers to monitor it. Given the difficulty in recruiting, not to speak of the limited budget, the District should not pull officers off the street to monitor cameras.

Reason No. 3: Surveillance Cameras Undermine Individual Privacy

Surveillance cameras like other post-September 11th measures exact an unacceptable toll on our individual liberties. Congresswoman Eleanor Holmes

Norton (D-DC) put the case for civil liberties very well. At a press conference on January 29th, to talk about the Open Society with Security Act, the Congresswoman said:

> September 11th accelerated the reach for blunt and often ineffective remedies to assure our security. Refusal of access through barricades and shut downs are everywhere in the country and we have now moved into the dangerous territory of infringements of personal liberty that has always enjoyed the highest protection . . . Intrusions into freedom of movement, privacy and our tradition of an open society are pervasive.

Surveillance cameras seriously intrude on individuals' right to privacy and have the potential to track individuals in their daily routines. In Britain, local police departments are already considering the use of facial recognition technology as a means to cross check individuals on the street against government databases. Other new surveillance technologies would pose increased threats to privacy under the Fourth Amendment. Thermal imaging technology would enable cameras to capture images from behind closed doors and night vision can bring images in complete darkness up to daylight level.

To get a glimpse of the future if the District were to proceed to create a British-style, surveillance camera system, consider the impact of public closed circuit TVs on the quality of life in Britain. George Washington University Professor Jeffrey Rosen provided the following account in the *New York Times Magazine* of October 7, 2001:

> Britain's experience under the watchful eye of the CCTV cameras is a vision of what Americans can expect if we choose to go down the same road in our efforts to achieve "homeland security." Although the cameras in Britain were initially justified as a way of combating terrorism, they soon came to serve a very different function. The cameras are designed not to produce arrests but to make people feel that they are being watched at all times. Instead of keeping terrorists off planes, biometric surveillance is being used to keep punks out of shopping malls. The people behind the live video screens are zooming in on unconventional behavior in public that in fact has nothing to do with terrorism. And rather than thwarting serious crime, the cameras are being used to enforce social conformity in ways that Americans may prefer to avoid.

Not only do surveillance cameras enforce social conformity as we learn from the British experience, but they inhibit protected First Amendment activity. We can expect that people will be much more reluctant to demonstrate on the Washington Mall and elsewhere in our nation's capital if they know that their images will end up in police files.

Reason No. 4: Permission to Establish Surveillance Cameras in the District Has Not Been Given

Americans value the right to be anonymous in a big city. We value the right to go about our business without the sense that the government is watching us as if we were going to break the law. All that would change if we were to import the British surveillance system. Before such a significant change can even be entertained, permission must be sought from those affected. That was not done in the District of Columbia. Neither the people nor their elected representatives on the Council were consulted. Instead, we read in the *Wall Street Journal* of February 13th that the plans for surveillance cameras in Washington "go far beyond what is in use in other American cities." The recent statements of the Mayor and the Chief of Police confirm that.

Reason No. 5: Surveillance Cameras are Subject to Great Abuse

Surveillance camera technology is readymade for abuse. The digitized images captured by the surveillance cameras are displayed on computer monitors at MPD headquarters, where federal law enforcement officials work with MPD officials. These images can be instantly retransmitted to other police agencies, which in turn can send them elsewhere. At any point, these images can be stored for future use.

While surveillance cameras are free of racial, gender, ethnic, and other biases, those who operate them may not be. We are able to monitor profiling by police officers when they make traffic stops. We will not be able to adequately monitor how officers select their targets for close observation by surveillance cameras. According to the University of Hull in Britain, "Black people were between one-and-a half and two-and-a half times more likely to be surveilled than one would expect from their presence in the population."

And we also know from the British experience that those monitoring the screens are likely to engage in CCTV voyeurism. Attractive women and romantic couples are special targets. In the same New York Times Magazine article, Jeffrey Rosen described the surveillance control room in Hull, England:

> During my time in the control room, from 9 p.m. to midnight, I experienced firsthand a phenomenon that critics of CCTV surveillance have often described: when you put a group of bored, unsupervised men in front of live video screens and allow them to zoom in on whatever happens to catch their eyes, they tend to spend a fair amount of time leering at women. "What catches the eye is groups of young men and attractive, young women," I was told by Clive Norris, the Hull criminologist. "It's what we call a sense of the obvious." There are plenty of stories of video voyeurism: a control room in the Midlands, for example, took close-up

shots of women with large breasts and taped them up on the walls. In Hull, this temptation is magnified by the fact that part of the operators' job is to keep an eye on prostitutes. As it got late, though, there weren't enough prostitutes to keep us entertained, so we kept ourselves awake by scanning the streets in search of the purely consensual activities of boyfriends and girlfriends making out in cars. "She had her legs wrapped around his waist a minute ago," one of the operators said appreciatively as we watched two teenagers go at it. "You'll be able to do an article on how reserved the British are, won't you?" he joked. Norris also found that operators, in addition to focusing on attractive young women, tend to focus on young men, especially those with dark skin. And those young men know they are being watched: CCTV is far less popular among black men than among British men as a whole. In Hull and elsewhere, rather than eliminating prejudicial surveillance and racial profiling, CCTV surveillance has tended to amplify it.

In addition, cameras provide new opportunities for privacy violations. Consider how a former MPD Lieutenant, Jeffrey S. Stowe, used his position to extort money from men who frequented gay bars in the District. How much better he could have blackmailed them if he had videotapes.

Conclusion

The District of Columbia should abandon its plans to establish a British style system of surveillance cameras. There are five reasons that compel this conclusion:

1. Surveillance cameras are not effective at fighting crime.

2. Surveillance cameras reduce resources for placing police officers into neighborhoods where they are needed.

3. Surveillance cameras undermine individual privacy and are inimical to the American way of life.

4. Surveillance cameras should not be contemplated without obtaining the explicit permission of those they impact. Permission was not granted in the District of Columbia.

5. Surveillance cameras are subject to great abuse.

Mrs. MORELLA. Thank you very much, Mr. Barnes. Very well presented. I am now pleased to recognize Ronald Goldstock of the American Bar Association. Thank you for being here, Mr. Goldstock.

Mr. GOLDSTOCK. I am the chair of the American Bar Association Criminal Justice Section Standards Committee and past chair of the section. I am pleased to have the opportunity to share with you the fruit of our labors from the ABA Criminal Justice Standards Committee. Typically, ABA standards, which are ABA policy, have been used extensively by courts and practitioners, and we are particularly pleased with its use in establishing legislative and regulatory policy.

The task force on technologically assisted physical surveillance [TAPS] that developed the standards we're discussing today was, I believe composed of the type of people that Ms. Norton suggested be on her proposed national commission. They included prosecutors and defense attorneys, academics, members of the judiciary, police, privacy advocates, and liaisons from a number of interested entities who worked very closely with the committee, entities like the FBI, NSA, NACDL, major chiefs of police, NLADA, and the Individual Rights and Responsibilities Section of the ABA.

The task force considered a variety of issues—"illumination and telescopic devices" from flashlights to satellite surveillance, "detection devices" from heat emanating from homes to x-rays at airports and "tracking devices" from car beepers to implanted chips.

We also looked very carefully at video surveillance, and in doing so, tried not to limit our analysis to present technology. We considered that anything that we could think of was possible, and we devised standards with that in mind. The standards consider a variety of possibilities.

If the video surveillance is of private activity or conditions, then we would require a warrant. Indeed, we would go beyond the Constitutional requirement of a search warrant, and require an eavesdropping warrant in such cases.

If the surveillance were to be long term overt surveillance, I think primarily the type that you're talking about today, then the decision to do so would have to be made by a politically accountable law enforcement officer or another relevant politically accountable individual—the Mayor, for example.

If the purpose of that surveillance, long-term overt surveillance, were investigative, then there would need to be a determination by that politically accountable individual that the surveillance would be likely to achieve a law enforcement purpose.

If, on the other hand, the purpose of such surveillance were merely deterrence, then it would require that notice be given prior to the time that the surveillance was installed, and notice be given while the surveillance was operating, and that there would be hearings before and during the surveillance so that the public would have a chance to comment, make known their views and a determination based upon those views.

Obviously, if it is going to be a deterrent, there would have to be notice, otherwise deterrence would not work. If the surveillance were short term and covert, then there would need to be a determination that a legitimate law enforcement objective could be met.

In such a case, the decision could be made under those circumstances, that is to say, short term, covert, by an officer who was not necessarily politically accountable because presumably the investigative demands would require that it be done immediately, it be done for a short period of time, but that it cease upon the attainment of its objectives.

We suggest all of this in the context of general principles, that there be no targeting in a discriminatory fashion, that there be an attempt to use less intrusive means before video surveillance is used, and that transactional data, the recordings themselves, be destroyed pursuant to some policy that was developed by the law enforcement authorities and the political structure, so that there wouldn't be the type of abuses that Johnny Barnes suggested. Thank you.

[The prepared statement of Mr. Goldstock follows:]

AMERICAN BAR ASSOCIATION

GOVERNMENTAL AFFAIRS OFFICE • 740 FIFTEENTH STREET, NW • WASHINGTON, DC 20005-1022 • (202) 662-1760

STATEMENT OF

RONALD GOLDSTOCK

on behalf of

THE AMERICAN BAR ASSOCIATION

before the

Subcommittee on the District of Columbia

of the

COMMITTEE ON GOVERNMENT REFORM
U.S. HOUSE OF REPRESENTATIVES

on the subject of

VIDEO SURVEILLANCE IN THE DISTRICT OF COLUMBIA

March 22, 2002

Madam Chair and Members of the Subcommittee:

My name is Ronald Goldstock. I am appearing on behalf of the American Bar Association at the request of its President, Robert E. Hirshon. The American Bar Association, the world's largest voluntary professional organization with more than 400,000 members, is the national representative of the legal profession. I appreciate the opportunity to appear before you to voice the Association's views on the use of video surveillance cameras.

Since August, 2000, I have served as Chair of the American Bar Association's Criminal Justice Section Standards Committee. I previously had served as a Committee member and as a Criminal Justice liaison to the Committee's Task Force on Law Enforcement and Technology. I am also a former Chair of the ABA's Criminal Justice Section.

The mandate of the Standards Committee is to continually review, recommend revisions to, and implement the ABA Criminal Justice Standards. The Standards are a multi-volume set of recommendations for criminal justice policymakers and practitioners. They span the entire spectrum of the criminal justice process, from the *Urban Police Function through Legal Status of Prisoners*. When the seventeen-volume first edition was published in the late 1960's, Chief Justice Warren Burger hailed the effort as "the single most comprehensive and probably the most monumental undertaking in the field of criminal justice ever attempted by the American legal profession in our national history." In the years since, several new volumes have been added, a second edition has been completed, and a third edition is well underway.

The first and second editions of the Standards contained detailed guidelines for conducting electronic eavesdropping of communications. The drafters of those Standards had considered including guidelines for the use of video surveillance and other technological means of solving and preventing crime, but had concluded that such standards would be premature. In 1995, the Committee decided that the time for such standards had come. It was clear that law enforcement use of technologically-assisted physical surveillance was increasing both in scope and complexity. It was also clear that there was little guidance from the courts or elsewhere, not simply in terms of specific rules but also with respect to the competing values to be weighed in making decisions about how and when to use these technologies in law enforcement. Accordingly, the Standards Committee created the Task Force on Technology and Law Enforcement. The charge to the Task Force was twofold: (1) to suggest revisions to the second edition Standards on electronic surveillance of private communications and (2) to develop proposed guidelines for the use of other "technologically-assisted physical surveillance." The Standards resulting from that second charge are the subject of my testimony today.

The Task Force was comprised of prosecutors, criminal defense lawyers, law enforcement officials, judges, privacy experts, and academics. In its efforts to identify and assess the primary constitutional and policy issues raised by the use of technologically-assisted physical surveillance, it consulted scores of organizations, ranging from national law enforcement agencies and local police departments to technology experts and advocates for individual privacy.

Following a two-year drafting effort, the proposed Standards underwent vigorous review by the Standards Committee, the Criminal Justice Section Council, and ultimately the ABA House of Delegates. They were approved as ABA policy in August 1998.

The technologically-assisted physical surveillance standards – the TAPS Standards, as they came to be called -- addressed tracking devices, illumination devices, telescopic devices, detection devices and video surveillance. I urge the subcommittee to review our printed publication for the detailed recommendations that I will only briefly describe here. The sections on the General Principles and on Video Surveillance are extremely relevant to the issues before this subcommittee and are appended to the subcommittee members' copies of my written testimony.

The TAPS Standards are grounded in the competing governmental and individual interests implicated by technologically-assisted physical surveillance. On the one hand, the Standards recognize such surveillance can be an important law enforcement tool. It can facilitate the detection, investigation, prevention and deterrence of crime, the safety of citizens and officers, the apprehension and prosecution of criminals, and the protection of the innocent. The Standards therefore recognize the need to allow such surveillance.

On the other hand, technologically-assisted surveillance can adversely affect individuals' privacy. While the most fundamental aspects of privacy are protected by the Fourth Amendment, there are other highly-valued aspects – including the general expectation of public anonymity -- that are not. As the economics of "surveillance by gadget" get cheaper and cheaper and the techniques themselves become increasingly sophisticated, the temptation is to use them more and more. Consequently, there is a very real danger that, unregulated, they can led to a stifling police presence affecting the innocent and guilty alike. The Standards therefore call for balancing the needs of law enforcement with the privacy interests of individuals to determine when the technological surveillance is warranted and the extent to which it should be regulated.

As noted above, there are a number of legitimate law enforcement interests in technological surveillance. In determining the weight to be given to them in any given case, the particular interests should be specified and a determination made as to the extent the surveillance would be likely to further them. The nature and extent of the crime to be detected or deterred and the harm to be protected against by the surveillance should also be assessed.

Privacy interests to be weighed include the nature of the place, activity, condition or location to be surveilled. There are also privacy interests in the availability and sophistication of the technology and the extent to which it enhances the natural senses, as well as in the physical scope of the surveillance. Whether the proposed surveillance is overt or covert is also an important consideration.

If the interests of law enforcement outweigh privacy interests, the Standards allow the use of video and other technologically-assisted physical surveillance subject to certain considerations. The Standards, for example, prohibit selecting the subjects of surveillance in an arbitrary or discriminatory manner. They require that any particular surveillance technique be capable of doing what it purports to do, and be used by trained officers. Further, they require that the technological surveillance be conducted solely for specified objectives and be terminated when those objectives are achieved. Appropriate notice is required. If the objective of the surveillance is deterrence, this entails advance notice so that people who do not want to be subjected to the surveillance can avoid it. If there has been a court order authorizing surveillance, it entails post-surveillance notice. The Standards also require that information obtained by the surveillance be used only for designated lawful purposes, and that protocols be developed for maintaining and disposing of surveillance records. Finally, the Standards require law enforcement agencies to prepare written instructions and train their officers about the requirements and rules of implementing various types of surveillance.

The Standards distinguish between video surveillance of private activities and long-term public video surveillance. While the former generally requires a judicially-authorized warrant, the latter is permitted when a politically accountable law enforcement official or governmental authority concludes that the surveillance will not view a private activity and will be reasonably likely to achieve a legitimate law enforcement objective. If deterrence rather than investigation is the primary objective, the public should be notified of the intended location and general capability of the camera and must have the opportunity, both prior to the initiation of the surveillance and periodically during it, to express its views. Covert and "short-term" overt video surveillance of public places are generally permissible if they will not view a private activity and are reasonably likely to achieve a legitimate law enforcement objective.

Finally, the Standards recognize the importance of holding government officials accountable for the use of technologically-assisted physical surveillance technology. Accountablity is a multi-pronged effort that includes administrative rules, using the exclusionary sanction when, and only when, it is mandated by federal or state constitutions or legislation; internal regulations; periodic review by law enforcement agencies of the scope and effectiveness of the surveillance, and providing the public general information about the type of surveillance being used and the frequency of its use.

Conclusion

The very issues your subcommittee is examining today have been examined in great depth by an ABA Criminal Justice Section Task Force on Law Enforcement and Technology. Comprised of a wide range of criminal justice professionals and privacy experts, the Task Force came to consensus on a detailed set of recommendations that subsequently underwent a rigorous review by the Criminal Justice Section Standards Committee, the Criminal Justice Section Council and the ABA policymaking House of Delegates. In the interest of the District of Columbia residents, visitors, and law enforcement officials, I urge your careful attention to the

*ABA Standards on Technologically-Assisted Physical Surveillance** as consideration is given to establishing what was described in your letter inviting the ABA to participate in this hearing as "the largest network of surveillance cameras in the United States."

Thank you.

*"The "black letter" Standards are available on-line at: www.abanet.org/crimjust/standards

Mrs. MORELLA. Thank you, Mr. Goldstock. And as I mentioned, as I introduced this panel, we have alluded to your ABA standards in our previous panel.

Mr. GOLDSTOCK. I assume that the standards, either in black letter or completely will be made part of the record.

Mrs. MORELLA. Indeed they will. Without objection, so ordered.

Mr. John Woodward, Jr., of RAND. Thank you for being with us, Mr. Woodward.

Mr. WOODWARD. Thank you, Chairwoman Morella, Ranking Member Norton, and members of the Subcommittee on the District of Columbia. I am honored to participate in these timely hearings to discuss electronic surveillance. In the interest of time I will focus my testimony on policy options for Congress to consider and concerns related to future use of video surveillance and other technologies.

Let's briefly establish the legal status quo by asking does the use of this technology violate legally protected privacy rights? The District of Columbia's proposed use of video cameras to monitor public places does not appear to run afoul of the protections afforded by the U.S. Constitution, specifically the unreasonable search provision of the fourth amendment. Congress, of course, is free to provide greater privacy protections than those found in the Constitution.

Thus we next need to ask what options can Congress pursue with respect to government use of surveillance cameras. It seems Congress has three broad policy choices: First, prohibit the use; second, regulate; third, take no immediate action pending further study. Congress could decide that the use of surveillance cameras is so dangerous to citizens in society that the law should prohibit it. In other words, the law could treat surveillance cameras as a form of technological heroin to be outlawed.

The counter argument is that surveillance cameras are not a form of technological heroin, because unlike heroin their use benefits the individual and the community.

Specifically, the community must maintain public safety. To fulfill this essential duty, law enforcement must monitor public places. This age old concept is the rationale for the police officer on the beat. To extend the analogy, a surveillance camera can be viewed as a form of mechanical police officer that watches or records events occurring in public places, places in which a person has no reasonable expectation of privacy.

Congress could decide to regulate the use of surveillance cameras in many ways. Most notably Congress could enact time, place or matter restrictions. For example, Congress could restrict government's use of surveillance cameras to the duration of a planned protest, time; to monitor only those sensitive locations deemed susceptible to attack, place; and to prohibit continuous video recording, manner. Or, as an alternative, Congress could require prior judicial branch approval for the technologies used.

The third option is for Congress to continue studying this issue, remaining poised to act. More and better public policy research on the use of the technology, its security and privacy implications as well as the effects of regulation is needed.

One controversial issue that Congress may consider concerns what the government does with the data it collects. The recording capability can be helpful for criminal investigations in the form of post event analysis. For example, it is good that we have a videotape of Mohammad Atta at the screening of the Portland, ME airport at 5:45 a.m., on September 11, 2001. It is bad that the Washington Metro CCTV system did not have a recording capability on June 10, 2001, when an assailant fatally shot Metropolitan Transit Officer Marlon Morales at the U Street Station.

If surveillance data are recorded, however, then policies for data retention, data security and auditing, along with penalties for misuse should be considered. Transparency and active oversight should help build citizen support.

Both the public and private sectors are making growing use of electronic surveillance technologies, as well as other emerging technologies that can gather information about us, thereby significantly increasing the potential privacy invasions.

These developments do not necessarily mean that big brother is alive and well. Rather, those same technologies can be used to protect citizens' rights; for example, with respect to the use of video cameras. In 1991, a bystander videotaped the arrest of Rodney King. This video allowed many Americans to see for themselves what otherwise would not have been seen.

Currently many police departments have installed cameras on squad cars to video certain traffic stops and similar events to show that officers satisfy legal requirements. In the near future, a citizen wearing a mini camera on her jacket lapel may be able to effortlessly video her encounters with law enforcement or anyone else and wirelessly transmit the information to wherever she desires.

In conclusion, we should be mindful of technologies that may invade our privacy. It is wise to monitor their development to forestall potential abuses. We should also, however, ensure that perceived or potential threats to our privacy do not blind us to the positive uses of video surveillance. Rather than acting hastily, it might be better to first try to develop thorough answers to questions about the use of surveillance cameras, as this hearing has tried to do, to determine the cameras' impact on public safety as well as the policy and social concerns they raise.

Thank you for inviting me to share my thoughts with you. I am happy to answer any questions you may have.

[The prepared statement of Mr. Woodward follows:]

"PRIVACY VS. SECURITY: ELECTRONIC SURVEILLANCE IN THE NATION'S CAPITAL"

WRITTEN TESTIMONY OF
JOHN D. WOODWARD, JR., ESQ.
SENIOR POLICY ANALYST, RAND

BEFORE THE
SUBCOMMITTEE ON THE DISTRICT OF COLUMBIA
OF THE COMMITTEE ON GOVERNMENT REFORM
UNITED STATES HOUSE OF REPRESENTATIVES

MARCH 22, 2002
WASHINGTON, D.C.

RAND is a nonprofit institution that helps improve policy and decisionmaking through research and analysis. This testimony is based on a variety of sources, including research conducted at RAND. The author thanks Professor Julie R. O'Sullivan of Georgetown University Law Center, Shirley Woodward, Esq. and Paul Steinberg for their comments. However, the opinions and conclusions expressed are solely those of the author and should not be interpreted as representing those of RAND or any of the agencies or others sponsoring its research.

Introduction

Chairwoman Morella, Ranking Member Norton, and Members of the Subcommittee on the District of Columbia, I am honored to participate in these timely hearings to discuss electronic surveillance. This forum provides an excellent opportunity for policymakers to examine government's use of such technologies as well as to gain greater understanding of the privacy and security concerns related to such use.

The many subject-matter experts and public officials appearing before the Subcommittee are better qualified than I to discuss the specific facts and the detailed history of electronic video surveillance in the District of Columbia. Therefore, I will focus my testimony on reasons for and concerns about the use of such technology, the legal status quo with respect to such use, options for Congress to consider, and issues related to future use of electronic surveillance and other technologies.

Reasons for and Concerns about Community Use of Electronic Surveillance

Over a million video surveillance cameras are used in the United States, with many government organizations deploying forms of electronic surveillance (*e.g.*, closed circuit television ("CCTV"), open circuit television ("OCTV"), video cameras, etc.) to monitor public places.[1] The United Kingdom has an estimated one million CCTV cameras in operation.[2] Various reasons have been cited for this extensive public sector use of this technology. These reasons include:

1. Preventing crime from occurring,
2. Detecting crime as it occurs,
3. Reducing citizens' fear of crime,
4. Aiding criminal investigations after a crime has occurred by what is known as post event analysis,
5. Making better use of public safety resources, and,
6. Countering terrorism.

Despite these reasons for using the technology, critics claim the use of surveillance cameras is not in keeping with "the values of an open society."[3] Some argue that their use leads to "video voyeurism" on the part of the police and forces "social conformity" on people. Others believe that the cameras' use may lead to racial discrimination.[4] For many, the government's use of surveillance cameras suggests that "Big Brother" is always out there secretly watching a citizen's every move. The Orwellian concerns are largely two-fold: the fear of tracking and the problem of clandestine capture of personal data.

Tracking refers to the ability to monitor an individual's actions in real time or over a period of time. In its most extreme incarnation, tracking could become a kind of "super surveillance" that lets the tracker to both "follow" a person today as well as search databases to learn where he was months ago.[5] The possibility that the government could compile such massive databases, and that such databases could be used by law enforcement, raises the specter of "Big Brother" tracking its citizens' every move.

The clandestine capture of a person's face increases these fears because surveillance cameras can surreptitiously track individuals without their knowledge or permission. Moreover, the information from tracking can be combined with other personal data, acquired by other means, such as credit card or other consumer purchasing records, to provide even more insight into an individual's private life.

Legal Status Quo of Community Use of Surveillance Cameras

Let's establish the legal status quo by asking: Does the use of this technology violate legally protected privacy rights? Legal rights to privacy may generally be found in three sources:

1. Federal and state constitutions (if the entity invading an individual's rights is a government actor),
2. The common law of torts (if the entity invading an individual's rights is a private actor), and,
3. Statutory law (at the federal and state level).

Although the word "privacy" does not appear in the U.S. Constitution, the concern with protecting citizens against government intrusions into their private sphere is reflected in many of its provisions. For example, the First Amendment protects freedom of expression and association, as well as the free exercise of religion, the Third Amendment prohibits the quartering of soldiers in one's home, the Fourth Amendment protects against unreasonable searches and seizures, the Fifth Amendment protects against self-incrimination, and the Due Process Clauses of the Fifth and Fourteenth Amendment protects certain fundamental "personal decisions relating to marriage, procreation, contraception, family relationship, child rearing, and education."[6]

Therefore, the constitutional "right to privacy" reflects concerns not only for one's physical privacy—the idea that government agents cannot barge into one's home—but also for less tangible interests—the idea that citizens should be able to control certain information about themselves and to make certain decisions free of government compulsion. And the Supreme Court has cautioned that it is "not unaware of the threat to privacy implicit in the accumulation of vast amounts of personal information in computerized data banks or other massive government files."[7]

A community's use of surveillance cameras could potentially implicate both types of privacy interests. For example, if the compilation of information in these databases had a significant chilling effect on First Amendment rights, such as discouraging citizens from attending political rallies, if it impinged on fundamental rights of decisional privacy, or if the information were insufficiently safeguarded against unauthorized disclosure, then the maintenance of such databases could potentially run afoul of the law.

Nevertheless, law enforcement's proposed use of the technology in Washington, D.C. does not appear to run afoul of the protections afforded by the U.S. Constitution.[8] Some civil libertarians may argue that a community's use of mass, dragnet video surveillance is improper, and that law enforcement should have individualized, reasonable suspicion that criminal activity is afoot before it can "search" a subject's face to see if the person is a criminal or terrorist. However, this argument is premised on what would be a dramatic expansion of current law. The Supreme Court has explained that government action constitutes a search requiring conformity with the Fourth Amendment when it invades a person's reasonable expectation of privacy. But the Court has also found that a person does not have a reasonable expectation of privacy with regard to physical characteristics that are constantly exposed to the public, such as one's facial features, voice, and handwriting.[9]

Similarly, the Court has determined that "[a] person traveling in an automobile on public thoroughfares has no reasonable expectation of privacy in his movements from one place to another." The Court reasoned "[w]hen [an individual] traveled over the public streets he voluntarily conveyed to anyone who wanted to look the fact that he was traveling over particular roads in a particular direction, and the fact of his final destination when he exited from public roads onto private property."[10]

So although the Fourth Amendment requires that a search conducted by government actors be "reasonable," which generally means that there must be some degree of suspicion that the person to be searched is engaged in wrongdoing, the government's taking video of a person's face that the person shows in a public place does not constitute a search.[11] And with respect to concerns about information privacy, if law enforcement officials limited their actions to simply comparing images of people in public places with databases of terrorists and known criminals, then information privacy concerns would be unlikely to arise.

Options for Congress

Congress is free to provide greater privacy protections than those found in the Constitution. Ample precedent exists for Congress to do so, particularly in the privacy realm. To cite one well-known example, In *United States v. Miller*, a case involving a bootlegger's private financial records, which were given to U.S. Treasury agents by his bank, the bootlegger tried to have this evidence excluded unsuccessfully.[12] The Supreme Court found that Miller had no expectation of privacy in the bank records, reasoning, "The depositor takes the risk, in revealing his affairs to another, that the information will be conveyed by that person to the Government. . .." The records could not, therefore, be considered confidential communications, because Miller had voluntarily conveyed them to the bank in the "ordinary course of business." Following the *Miller* decision, Congress, in 1976, passed the Right to Financial Privacy Act (12 U.S.C. 3401 *et seq.*) to establish procedural requirements that financial institutions must follow when disclosing information to the government. Thus, the refusal of the Supreme Court to find a Constitutionally-mandated privacy right does not as a practical matter end the debate. In sum, Congress is free to act by passing legislation to create new individual privacy rights.

Having established that Congress can act in this arena, we next need to ask: What options can Congress pursue with respect to government use of surveillance cameras? It seems Congress has three broad policy choices:

1. Prohibit its use,

2. Regulate its use, and,
3. Take no immediate action, pending further study.

1. Prohibit its use: Congress could decide that the use of surveillance cameras is so dangerous to citizens and society that the law should prohibit such use. In other words, the law could treat surveillance cameras as a form of "technological heroin" to be outlawed. In the process, Congress could help to establish an all-encompassing individual right to "public privacy," that is, regardless of whether the individual is in a private place (e.g., the home) or in a public setting (e.g., the street), the government cannot watch or record the event with surveillance cameras.

The counterargument to this approach is that surveillance cameras are not a form of "technological heroin" because, unlike heroin, their use benefits the individual and the community. Specifically, the community must maintain public safety. To fulfill this essential duty, law enforcement must monitor public places. This age-old concept is the rationale for the "police officer on the beat." To extend the analogy, a surveillance camera can be viewed as a form of mechanical police officer that watches or records events occurring in public places in which the person has no reasonable expectation of privacy.[13]

2. Regulate its use: Congress could decide to regulate the use of surveillance cameras in many ways.[14] Most notably, Congress could place statutory restrictions on use in the form of time, place or manner restrictions.[15] For example, Congress could restrict the District of Columbia's use of surveillance cameras to the duration of a planned protest (time), to monitor only those sensitive locations susceptible to attack (place), and to prohibit continuous video recording (manner).

Or as an alternative, Congress could require prior judicial (or executive) branch approval for the technology's use. A rough parallel can be drawn to the legal requirement for law enforcement to obtain a warrant prior to intercepting certain communications. Congress could extend a similar requirement to law enforcement's use of video surveillance. On this note, the House of Delegates of the Commonwealth of Virginia recently took this approach by passing House Bill 454, requiring that state and local law enforcement agencies obtain an order from a circuit court prior to using facial recognition technology.[16] The legislation has been sent to the Virginia State Senate.

3. Take no immediate action, pending further study: The third option is for Congress to take no immediate action in the nature of prohibition or regulation, but rather to continue studying and assessing the issue, and remain poised and prepared to act if deemed necessary or advisable. Based on what it has learned at this hearing and at similar public forums, Congress can begin to formulate more complete answers to questions about the use of the technology. As a start, these questions include:

1. Does electronic surveillance prevent crime? Or does it merely shift crime to other locations?
2. How effectively does electronic surveillance detect crime?
3. How do citizens perceive the technology's use? Do citizens favor the

use of the technology?
4. How much of an aid is electronic surveillance in criminal investigations?
5. Does electronic surveillance enable public safety resources to be deployed more efficiently and effectively?
6. Does electronic surveillance counter terrorism?

Moreover, Congress can be particularly attentive to any solid evidence of actual harm caused by the technology. For example, when use of the technology leads to misidentifications resulting in the arrest and conviction of innocent individuals. Granted, it is very hard to measure or to develop the "metrics" needed for these answers. More and better public policy research on the use of the technology, its security and privacy implications, as well as the effects of regulation is needed.

As it develops answers to these and related questions, Congress has a plethora of policy choices to consider. As with any technology, public understanding of the operation and uses of electronic surveillance may mitigate many of the fears about Big Brother. To that end, the government should be encouraged to use the technology openly, rather than clandestinely. To the extent possible, prominent notice should be given that the technology is being used. Moreover, the government entity using video surveillance should provide as much information as possible to the public about the technology's purposes and capabilities. This transparency should help prevent misuse or abuse. This effort at transparency could even include broadcasting the actual operations of the electronic surveillance on a public access television channel.[17]

One controversial issue that may arise concerns what the government does with the data it collects. Thus a first question to consider is whether the surveillance camera systems should be limited to actively monitoring or if they should have a recording capability as well. A recording capability can be helpful for post-event analysis. For example, it's good that we have videotape of Mohamed Atta at the screening area of the Portland, Maine airport at 5:45 a.m. on September 11, 2001. It's bad that the Washington Metro CCTV system did not have a recording capability on June 10, 2001, when an assailant fatally shot Metropolitan Transit Officer Marlon Morales at the U Street station.

If surveillance data are recorded, however, then a retention policy should be considered. After a certain time period, the data should be erased. Similarly, the data should be securely kept. And an auditing process should be established to ensure accountability and to prevent the misuse of the data.

Finally, some form of active oversight, either government-only or a cooperative effort between government officials and private citizens, such as citizen oversight committees, would be useful not only to quell fears about the technology's use but also to ensure that it will not be abused.

Future Use of the Technologies

These hearings are extremely timely, and this topic will continue to occupy the attention of policymakers because of continuing technological developments. Both the public and private sectors are making growing use of electronic surveillance technologies as well as other emerging technologies that can gather information about us.[18] We are also seeing increasing interlinkages among these various information-gathering technologies. The chart below shows these interlinkages.[19] The center of the chart depicts average citizens, all of whom have different types of personal information associated with them (e.g., biographical, consumer, biometric, medical, financial, and behavioral). Surrounding this information are various technologies that can obtain this personal information (e.g., overhead imaging, genomics, biometrics, thermal imaging, computer monitoring, data mining, micro sensors, global positioning systems, geographical information systems, surveillance cameras (CCTV), and bio-nano-materials technology). When these technologies are interlinked, the ability to collect information, and the amount of information collected increases significantly, thereby significantly increasing the potential privacy invasions.

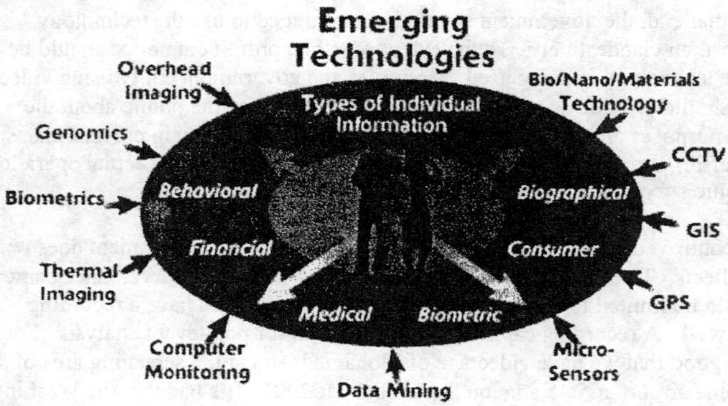

A well-publicized example of such interlinkages of the technologies occurred at Super Bowl XXXV at Raymond James Stadium in Tampa, Florida on January 25, 2001 when authorities used facial recognition technology and surveillance cameras to capture spectators' biometric information, combined with a little data mining, to determine biographical information (e.g., is this person a wanted criminal?). The chart below shows these interlinkages.

Interlinkages of Technologies

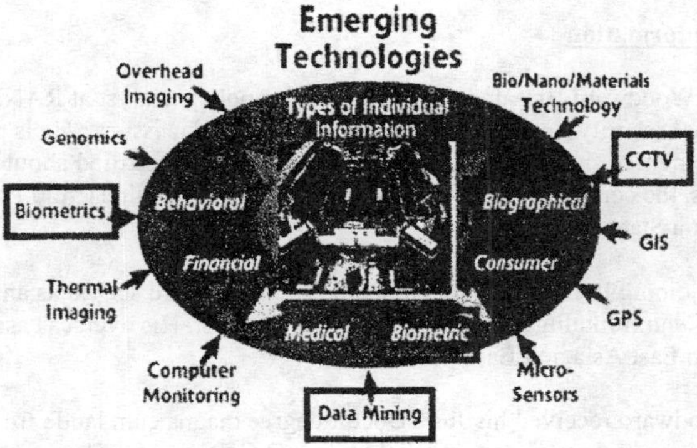

And this use of interlinked surveillance technology is not limited to law enforcement. We have now not only entered the world of webcam, Kindercam, nannycam, and grannycam, but we have also entered the realm of countless other means of information-gathering technologies. These developments do not mean that Big Brother is alive and well. Rather, these same technologies can be used to protect citizens' rights. For example, with respect to video cameras:

- In 1991, a bystander videoed the arrest of Rodney King; this video allowed many Americans to "see for themselves" what otherwise would not have been seen.
- Currently, many police departments have installed video cameras on squad cars to video certain traffic stops and similar events to show that officers satisfy legal requirements.
- In the near future, a citizen, wearing a mini-camera on her jacket lapel, may be able to effortlessly video her encounters with law enforcement or anyone else and wirelessly transmit the information to where she desires.

In conclusion, the Subcommittee on the District of Columbia is holding a hearing on an important topic of public concern that presents complicated privacy and security issues, and possibly balancing of the two. Moreover, the technology moves quickly and the law and policy concerns often struggle to keep apace. However, this situation is not new in the American experience. We should be mindful of technologies that may invade our privacy, and it is wise to monitor their development to forestall potential abuses. We should, however, also ensure that perceived or potential threats to our privacy do not blind us to the positive uses of electronic surveillance. And rather than acting hastily, it might be better to first try to develop thorough answers to questions about the use of surveillance cameras, their impact on public safety, and the policy and social concerns they raise. Thank you for inviting me to share my thoughts with you. I am happy to answer any questions you may have.

Biographical Information

John D. Woodward, Jr., an attorney, is a senior policy analyst at RAND where he works on national security, intelligence, and technology policy issues. He is particularly interested in biometrics and surveillance technologies. He has testified about biometrics before Congress, the congressionally-created Commission on Online Child Protection, and the California State Assembly.

Prior to joining RAND full-time in 2000, Mr. Woodward served as an Operations Officer for the Central Intelligence Agency for twelve years. His overseas assignments included tours in East Asia and East Africa.

Mr. Woodward received his Juris Doctor degree magna cum laude from Georgetown University Law Center in Washington, D.C. He was a Thouron Scholar at the London School of Economics, University of London, where he received his M.S. in Economics. He received his B.S. in Economics from the Wharton School of the University of Pennsylvania. He served as a law clerk to the Hon. Roderick R. McKelvie, a U.S. District Court Judge in Wilmington, Delaware.

---

[1] Closed circuit television ("CCTV") refers to a system in which the circuit is closed and all the elements are directly connected. Open circuit television ("OCTV") transmits video images over wireless networks to portable devices. Cameras are used for both systems.

[2] *See, e.g.*, Parliamentary Office of Science and Technology (United Kingdom), "Biometrics and Security," *postnote*, Nov. 2001 at 2. The Parliamentary Office of Science and Technology will soon publish a *postnote* discussing the use of surveillance cameras in the United Kingdom.

[3] Jeffrey Rosen, "A Cautionary Tale for a New Age of Surveillance," *New York Times Magazine*, Oct. 7, 2001. Rosen further states "They [surveillance cameras] are technologies of classification and exclusion. They are ways of putting people in their place, of deciding who gets in and who stays out, of limiting people's movement and restricting their opportunities."

[4] According to Barry Steinhardt, Associate Director of the American Civil Liberties Union, "People of color in the U.K. have been singled out to be followed by the cameras." quoted in Stephen Cass and Michael J. Riezenman, "Improving Security, Preserving Privacy," *IEEE Spectrum Online*, Jan. 4, 2002. In his *New York Times Magazine* article, Jeffrey Rosen states "In Hull [United Kingdom] and elsewhere, rather than eliminating prejudicial surveillance and racial profiling, CCTV surveillance has tended to amplify it."

[5] For a more complete discussion, see John D. Woodward, Jr., *Super Bowl Surveillance: Facing Up to Biometrics*, RAND Issue Paper 209, 2001.

[6] *Planned Parenthood of Southeastern Pennsylvania v. Casey*, 505 U.S. 833, 851 (1992).

[7] *Whalen* v. *Roe*, 429 U.S. 589, 605 (1977).

[8] States are free to provide greater privacy protections in their own state constitutions than those afforded in the U.S. Constitution. When evaluating the use of a specific system, therefore, its legality must be analyzed under state constitutional provisions as well.

[9] *United States* v. *Dionisio*, 410 U.S. 1, 14 (1973).

[10] *United States v. Knotts*, 368 U.S. 276, 281-82 (1983).

[11] *See also* U.S. Department of Justice, *Criminal Resource Manual* 32 (Oct. 1997) ("Video surveillance, which is the use of closed-circuit television (CCTV) to conduct a visual surveillance of a person or a place, is not covered by Title III. Rather, its use is governed by the Fourth Amendment and, therefore, when a reasonable expectation of privacy exists, a search warrant should be sought.")

[12] *United States v. Miller*, 425 U.S. 435 (1976).

[13] *See, e.g.*, Marcus Nieto, *Public Video Surveillance: Is It an Effective Crime Prevention Tool?* CRB-97-005, (California Research Bureau, California State Library, June 1997) at 4 (footnotes omitted).

[14] The United Kingdom, for example, regulates the public and private sector use of CCTV systems with a CCTV Code of Practice issued under the Data Protection Act 1998. *See, e.g.*, Office of Information Commissioner available at http://www.dataprotection.gov.uk/.

[15] In October 2001, the Privacy Commissioner of Canada, George Radwanski, released a letter of finding following his investigation of video surveillance activities by the Royal Canadian Mounted Police (RCMP) in Kelowna, British Columbia. In his letter of finding, Commissioner Radwanski suggested several such limitations be placed on the RCMP's use of video surveillance, based on his interpretation of Canada's Privacy Act. His suggestions include limiting the use of video surveillance, without continuous recording, to "particularly sensitive locations that are so susceptible to some form of terrorist or other attack as to require intensive security measures," a "particular place [where] there might be such an exceptional threat to public safety, combined with other circumstances that made conventional policing unfeasible," and "special circumstances where, to investigate a particular crime, it might be appropriate for police to temporarily establish a video camera in a given location and record images of everyone who frequents that location." *See* Office of the Privacy Commissioner of Canada, *News Release: Privacy Commissioner Releases Finding on Video Surveillance by RCMP in Kelowna*, Oct. 4, 2001 available at http://www.privcom.gc.ca/media/nr-c/02_05_b_011004_e.pdf. On March 15, 2002, Commissioner Radwanski sent a letter to Lawrence MacAulay, Solicitor General of Canada, regarding the RCMP's continued use of video surveillance in Kelowna. *See* Office of the Privacy Commissioner of Canada, *News Release*, Mar. 15, 2002 available at http://www.privcom.gc.ca/media/nr-c/02_05_b_020315_e.pdf

[16] *See House Bill 454*, dated Feb. 11, 2002 available at http://leg1.state.va.us/cgi-bin/legp504.exe?021+ful+HB454EH1+pdf.

[17] For a provocative discussion of surveillance and the need for transparency, *see* David Brin, *The Transparent Society: Will Technology Force Us to Choose Between Privacy and Freedom?* (Perseus, 1998).

[18] *See, e.g.*, Ivan Amato, "Big Brother Logs On," *Technology Review*, Sept. 2001 at 58-63.

[19] This chart grew out of discussions at a "Privacy and Emerging Technologies" Conference hosted by RAND in July 2001. My former RAND colleague, Kristina Larson, contributed immensely in the development of this chart.

Mrs. MORELLA. Thanks. I want to thank the three of you for appearing before us and so succinctly giving us a synopsis of your work on this issue.

I am going to ask some questions that would be directed actually to all of you. I am wondering, first of all, have any jurisdictions that have installed electronic surveillance systems adopted the standards that were recommended by the ABA? You are probably the first one to have a response to that, Mr. Goldstock.

Mr. GOLDSTOCK. Yes. I don't know of any.

Mrs. MORELLA. Anybody know of any? OK. When did you come out with them, as a matter of fact? I know it was—2 years was your task force, as I recall?

Mr. GOLDSTOCK. In 1999.

Mrs. MORELLA. 1999?

Mr. GOLDSTOCK. Yes. They came out toward the end of 1999.

Mrs. MORELLA. Great. Well, it is a very interesting that nobody knows about any jurisdiction that has utilized that work.

OK. Would you say that the recording of the images and the use of recorded images and data bases poses the greatest risk to civil liberty? Any of you want to comment on that? I mean, is having it in data bases a real threat?

Mr. BARNES. Absolutely, Madam Chair, because what we will be exposed to in that case is this idea of mission creep and how far will it go. Will persons who have done other things considered to be wrongs in society be excluded because they didn't pay a traffic ticket, or they didn't pay a judgment to the courts?

The collection of information and the storing of information about individuals is a very dangerous thing. Let me just say that I know of an example of a bank that got the information, the health information about their patrons and those who were suffering from cancer. They called in their notes. In other words, you are about to die, give us our money. So there are many abuses that flow from the storing—the collection and storing of this kind of data.

Mrs. MORELLA. Mr. Woodward.

Mr. WOODWARD. Chairwoman, if I may followup. Is the recording of the images and storing them in a data base the greatest risk? I am reluctant to give it an order of magnitude. I think it is certainly a risk that deserves the attention of Members of Congress.

I think Mr. Barnes' concern about function creep or mission creep is well placed. That is sadly an example that we see from history with much use of data that is stored.

There frequently are other reasons found why that data can be used in ways never intended, with the Social Security number being a great example. I would also add, I think another concern that Members of Congress will face in the future, you will have this data. If you decide to store it, and there may be very, very good and compelling reasons why you want to store the data. For example, it may be very helpful for criminal investigation, post-event analysis work. At the same time you will see growing technologies that can also collect information about people, that might become interlinked in a type of surveillance network that I think could be profoundly troubling for many of us in society.

And I would also add that it is not the public sector that has this monopoly on the use of these technologies by any means. This is also something that you will see extensively in the private sector.

So I think you have rightly called this hearing to focus attention on this. I think that this will continue to be an issue that will occupy much of your attention as policymakers.

Mrs. MORELLA. What do you think about the 72 hours that Chief Ramsey mentioned for keeping that material?

Mr. WOODWARD. Ma'am, this is in reference to the Metropolitan Police Department's draft regulation that would call for a 72-hour storage period?

Mrs. MORELLA. Right.

Mr. WOODWARD. I think that there definitely needs to be a time limit as far as how long the data is going to be stored. I don't have enough specific knowledge of the situation that Chief Ramsey and his colleagues face as to why they came up with the 72-hour figure, but it seems to me a good start.

Mrs. MORELLA. OK. And you mentioned the—you both mentioned mission creep. And you mentioned technology coming on line so quickly. I mean, the idea of having a little camera on our lapel as we are out in the evening or in certain areas. How do we guard against it? What do we do to guard against it? Whoever would like to answer that. With so much new technology coming on line.

Mr. BARNES. Yes, Madam Chair. Again, I think we need a body of laws and a set of regulations just as we have with respect to search warrants, with sanctions and criminal sanctions and civil penalties for abuse. And there have been abuses, well documented, well reported. I think that is the best way.

The concern is that this has been a unilateral action by the executive branch cutting out the legislature, cutting out the courts. And the genius of our system of separation of branches is that we do have checks and balances. And that is why the courts and the legislature need to be involved. We need a body of laws, a set of regulations if we are to have those cameras at all.

Mr. GOLDSTOCK. It seems to me that we are talking about a number of different things and they have to be answered separately. One question is: What government can do, what they are allowed to do and whether or not that should be regulated by laws, by the Constitution, by policy, by regulations. All is a matter that we need to think about.

The second is government's access to private data bases. There are huge data bases that are built up both commercially and in other ways. Every time you use an Easy Pass to go through a toll or go into a parking lot there are records maintained that can be retrieved. We need to know what the government ought to be able to obtain and under what circumstances private companies or entities can share their data with law enforcement. Huge retailers ask you for your telephone number when you make a purchase, collect data about you, send you packages in the mail. They know where you live. All of that exists. Links can be made internally within those retail stores. They can be provided to law enforcement either on their own design or by request of law enforcement. We ought to be thinking about how we are going to regulate that.

Finally, it seems that there is the possibility of violations of law by individuals, whether or not it would be a crime, for example, for somebody to broadcast a picture or overhear a communication, and that obviously is subject to law. Currently eavesdropping laws regulate some of that; and there might otherwise be other laws as the capabilities and technology increases.

But I think these are incredibly complex issues that need to be studied, and to suggest that there is a simple answer I think is probably incorrect.

Mrs. MORELLA. OK.

Mr. WOODWARD. May I add to that, Chairwoman Morella? With respect to mission creep, I think it is important to point out that mission creep itself should be really viewed as a neutral term. That is, one person's mission creep might be another person's worthwhile mission advance.

And let me try to give you an example. In the testimony I heard earlier, it seems to me that the Metropolitan Police Department currently proposes using video surveillance for special events, special occasions, and the Chief gave examples.

Well, it might be that over time that instead of just using the surveillance on a limited basis, the community might want to use it more on a regular basis, normal daylight hours for regular standard law enforcement purposes. With the use of special cameras that have a night vision capability, you can see how that could be expanded to a 24/7 capability, that the surveillance could become wider, it could become deeper, etc.

But I think you might be faced with the question of what do you do when the community, be it the local community or at the State level, or at the national level, actually approves this use, and wants this kind of mission extension, if you will. And to give you an example, some of us might think, well, it is really silly to use video surveillance to deal with a problem we have of people who are smoking illegally in certain public areas and we are going to train the surveillance cameras on that area to try to identify those people so that we can prevent this practice and we can cite the people who are committing the infractions.

But it is difficult if you have a community that goes through public debate, engages in the democratic process and decides, well, in our community we have a real problem with people who want to smoke illegally in public areas and we want to stop it. It is difficult. Some might look at that and say, well, that is really a silly and unwarranted use of the technology. But to others who are exercising their democratic rights, they might see it as something they really want to stop. So I think it is a very difficult issue.

Mrs. MORELLA. Well, the difference is mission creep versus mission enhancement. It is all in what you call it. Great.

Ms. Norton.

Ms. NORTON. Mr. Woodard, when you get into the notion that the community can decide that it wants cameras for any purpose, there is a consensus, you and I would have a deep disagreement with that because there is a Constitution of the United States and I can imagine all kinds of things that there is a consensus about that would not pass constitutional muster, I think and I hope.

Mr. Barnes, it is clear as you opened your testimony that the ACLU is trying not to be a bunch of Luddites that simply oppose the cameras wherever they are found, that you recognize that the cameras exist, that the cameras for many uses have not been struck down by courts yet, that they have been challenged.

But you say that we should abandon—I am sorry, that the District should abandon plans for a British style system, which of course is a wholesale system. In light, by the way, of that opening in your testimony, let me indicate that your reference to profiling suggests to me that we are talking about exactly the opposite of profiling. Profiling is honed in on specific individuals because of certain characteristics. Here we are talking about wholesale profiling.

Everybody is a target. It is hard to know which is worse, for everybody or some of us. But I do see, as I listen to the testimony of all three of you, the willingness to try to work this through, to come to some—to recognize that it is difficult to come to some kind of solution that all could embrace.

Mr. Goldstock, my commission is a little broader. Your commission, I want to compliment the ABA. I didn't know of your work. I want to compliment the ABA for getting out in front of this issue. You seem to have been focused on surveillance cameras. The reason mine is broader is that I would hope that the Presidential commission obviously wouldn't be involved in regulations, would provide general guidance across the board on how to meet—how society remains open. It includes things well beyond surveillance, because there are so many approaches being thrown up.

Nevertheless, your work, I think it would even be important to what it is I would like to have done. I mean, my commission is going to have on it not only lawyers and security officials, but philosophers and historians and psychologists and architects, people who can bring to bear the society so that we can have some balance here. And your work will be important, it, seems to me, if the commission is ever established.

I would like to see whether or not there could be any agreement among those of you seated at the table on the framework for an accepted policy. Mr. Goldstock's testimony mentions technique capable of doing what it purports to do, trained officers solely for specified objectives, terminated when the objective is achieved, notice, deterrence, if there is deterrence, notice so that everybody knows what is happening as well, maintaining or disposing of those records, written instructions, etc.

I see something approaching the components of an acceptable policy. And if I could add to that, the testimony from the city about audits so that one could in fact see if in fact what is—what you said should happen is happening. No data base. Could I ask you whether the components I have just named are, as far as the three of you are concerned, the framework toward an acceptable policy? If not, what should be added? Or is there some other approach you would suggest?

Mr. GOLDSTOCK. I would say yes. I don't have any problem with it.

Ms. NORTON. I used yours because it laid out some. I added, of course, one or two from D.C. and I would like to ask Mr. Barnes, who I think very usefully raised these issues in our city, whether or not the ABA policy, plus some of the things that the District may be doing now that the matter is public, could form the framework for an acceptable national policy?

Mr. BARNES. Well, Congresswoman, as you note, our position is that we don't believe the cameras serve any useful purpose. But, as you also note, some cameras are in place. Others perhaps are contemplated. We would suggest that if there is to be some system, that a body of laws in a set of regulations, certainly those put forth by the American Bar Association reach many of the concerns that we have.

And we would be happy to work with others in trying to shape and mold a body of laws and set of regulations that reflect a concern about civil liberties.

Ms. NORTON. Yes. I appreciate that, that response. Mr. Woodward.

Mr. WOODWARD. Yes, Congresswoman Norton. I think you cited—I was trying to keep a checklist, and I think I got most of them. I think you mentioned that the legal framework should have something like notice, or people should have notice that these surveillance cameras are being used in a particular area.

Ms. NORTON. Yes. Two kinds of notice. Notice to let us know so that we can comment on whether they should be used and notice that they are there.

Mr. WOODWARD. Right. I think you need both, basically. I think it is very important to iron out your policies or your regulations that concern this whole issue of data retention. What you can keep in a data base. Similarly with data security, an auditing function as well as some kind of oversight mechanism.

I think also you mentioned the idea of penalties for misuse. I think it is appropriate for the Congress to look at criminal and civil penalties, because you certainly want to deter that kind of behavior. I think also the important aspect to this is transparency and getting input from the community as far as what is the right way to work this process.

Now, transparency of course is always difficult; how much transparency. But I know that I have considered many ideas, and I include in my written testimony the idea that the operations of these video surveillance cameras could be made known to the community by broadcasting some of their activities on a public access channel or on a Web site so people could literally see that this is what our government is doing in these areas.

Ms. NORTON. Thank you. My time is up. I only have a few more questions.

Mr. GOLDSTOCK. Let me just make one thing clear, That the ABA policy regarding notice, relates to overt long-term surveillance. If the surveillance were being used for an investigative purpose, if they knew there was goals to be a break-in to a liquor store, obviously you wouldn't be giving notice.

Ms. NORTON. That is with a warrant?

Mr. GOLDSTOCK. No. Without a warrant because it was on a public street but focused in on a particular area for investigative purposes.

Ms. NORTON. Probable cause, and therefore the——

Mr. GOLDSTOCK. No.

Ms. NORTON. The cop is not just there in front of somebody's liquor store. What do you mean?

Mr. GOLDSTOCK. What I mean is if the police, for example, had information that a group of burglars or robbers were breaking into liquor stores, and there weren't a sufficient number of cops to be around protecting everybody, you might want to set up cameras at a variety of stores in that neighborhood aimed directly at the stores. That would not require a warrant. But you wouldn't want to give notice because the object was investigative rather than deterrent.

Ms. NORTON. Would Mr. Barnes comment on that hypothetical?

Mr. BARNES. Well, if I may, Congresswoman, I feel obliged to say, because I may not get another opportunity, that I have worked with both of your staffs and you have excellent staffs.

I think the hypothetical is a specific situation involving a particular circumstance. And the trouble we have with the surveillance cameras is that they are general and sweeping. And I think that as a standard would also be important, to focus on particular situations, specific circumstances not unlike we do with Fourth Amendment procedures.

Ms. NORTON. Thank you.

Mrs. MORELLA. Thank you, Congresswoman Norton.

You all heard in the first panel that the Council has passed emergency legislation that requires that the Metropolitan Police Department submit policies and procedures governing the use of the surveillance and asking for approval.

Do you think that approach is going to be sufficient to ensure that there will be some regulation of the system, adequate regulation of the system, or is that just a beginning?

Mr. BARNES. Well, again, Madam Chair, we believe that the effort to put some controls and guidelines need to go far beyond the police issuing some statements about what it might do or will do. It needs to be etched in the law and have the force and effect of the law.

And there needs to be an opportunity for those who are victimized by abuses to have access to the courts to enforce that law and those regulations. So we need more than policy guidelines.

Mrs. MORELLA. OK. I know you want to comment on that. But maybe under the umbrella question, do I hear the three of you saying: We really do think that Congress should at some point come in with standards, policies, procedures, auditing devices, and sanctions?

Are you saying there should be legislation that is drafted to address what you have presented to us in terms of the problems that could be inherent with camera surveillance?

Mr. GOLDSTOCK. I think that the ABA standards could be the basis of legislation. I think it would work quite well. I think the legislation, as you suggest, is only the first step. In addition, there has to be training and transparency and auditability and accountability and oversight. I think—and I think I am in agreement with Johnny Barnes on this—that in fact the real issue here isn't the initial policy, although I think that is important. It is consistency. It is in demonstrating to the public that the policies are being adhered to, that there aren't abuses, that if there are issues that come up, they are attended to. That maybe there should be a change in legislation or policy if there are consistent abuses and that the legislation standards and regulations change to meet current needs and current technology.

I think that is what is critical. This is an ongoing process. It is not the beginning, formulating a manual and then stopping there. This is a continuing process and I think that is what is necessary.

Mrs. MORELLA. Should it be Federal, or should it be jurisdictional?

Mr. GOLDSTOCK. Well, I think the broad guidelines ought to be Federal. I think—the same way it is in electronic surveillance. But the electronic surveillance statutes, for example, lay down bare minimums, and different jurisdictions decided within that, they ought to make changes and be more restrictive, and I think that is the same kind of policy that would work here as well.

Mrs. MORELLA. So kind of bare minimums allowing for some flexibilities and customizing for different areas. Mr. Woodward.

Mr. WOODWARD. Yes, Chairwoman. I think you are at a starting point. You could look to Federal regulatory framework for at least how you want Federal agencies to use the technology. And then you might want to consider, should we also include State, local, tribal government use of the technology as well.

One other point I would urge you to consider, and I think we have seen it as far as the testimony presented this morning before you. It seems to me there is a lot of disagreement as to how effective surveillance cameras are in terms of preventing crime, detecting crime, investigating crime, and countering terrorism. It seems to me that we have heard different knowledgeable people offer different perspectives on that point.

It just seems that one area Congress might want to further investigate is to try to determine some answers. I realize it is very hard to develop the metrics so that you can get answers to questions. Well, is this technology really cost effective and so on? I do understand that there has been a lot of reference made to the United Kingdom's use of surveillance cameras by their government agencies. I just wanted to call the subcommittee's attention to the fact that the Parliamentary Office of Science and Technology, which is a very, very rough United Kingdom equivalent of your own Congressional Research Service, will be issuing a report in the very near future I understand on the topic of surveillance cameras.

Now, that might have some helpful insights as far as how the United Kingdom Parliament perceives the use of the technology in that particular case. And also I would just note from an international perspective that our neighbors to the north in Canada have taken a rather different approach in some ways to the use of surveillance cameras to monitor public places. Privacy Commissioner Radwanski has issued a letter. Although it doesn't have the force of law, he outlines his comments and his views of how the technology should and should not be used. And apparently there is a disagreement among various lawmakers in Canada as far as what the policy will be there concerning government use of surveillance cameras for public areas.

Mr. BARNES. May I just quickly add to that, Madam Chair? I want to join with Mr. Woodward. I think it would be useful to find out whether or not these systems have been effective in other places. Again, I know of a report that came out of Sydney, Australia, and Chief Ramsey referred to certain data from Sydney, Australia in December 2001 that indicated that these cameras had helped with one arrest every 160 days.

That is different than what the Chief told us. So it would be, I think, a useful exercise to find out the effectiveness of those cameras.

Mrs. MORELLA. How long have they had it in London? I don't remember——

Mr. BARNES. In England the system was first put in place following the IRA bombing in 1994. Second one in 1995. And then when the little 4-year-old was kidnapped it just proliferated. And Sydney, Australia, I am not sure how long they have had it.

Mrs. MORELLA. Thank you. Ms. Norton.

Ms. NORTON. Thank you, Madam Chair.

I think it was Mr. Goldstock that mentioned the necessity for consistency. I was troubled by even the Interior Department testimony. Mr. Parsons sensed that various Park Service facilities, that is only one department of the government, were essentially on a case-by-case basis.

We saw what happened in this city when Federal agencies proceeded on a case-by-case basis and barricades went up, all kinds of rights were violated. Parking meters off—when it came to some Federal agencies, not bothered with respect to others, violation of District law with respect to barricades.

The point of consistency does lead me to believe that some Federal action guidance of some kind is necessary. The case—we work brilliantly in this country on a case-by-case basis. But that assumes that there is a law or a basis by which to judge the case by case.

And how we go about that in—if we are the Federal Government, it becomes problematic. At the very least it does seem with me with respect to Federal facilities there is an obligation. We probably should begin there, because we are just learning, before we begin to tackle our national responsibilities.

Would you agree that if we were going to proceed in some kind of Federal policymaking, that we should proceed first with Federal facilities, and do you think that should be a matter of law or some kind of executive order or central guidance?

Mr. GOLDSTOCK. Well, it seems to me that the Federal policy, whether it be for Federal facilities or even for local, be minimum standards. It seems to me that is what can be done the best, to have a sense of consistency throughout the country and to put into place the kinds of policies and formulations that you agree exist after hearings and these studies.

But it does seem to me that there ought to be a great deal of flexibility with respect to individual incidents and the questions of whether or not there are particular needs in particular jurisdictions, either because law enforcement is not broad enough or hasn't enough people that they have to enhance their resources. There may be particular problems that exist.

I should also say that in considering this, it is not clear to me that it is always privacy versus law enforcement. Sometimes there can be an accommodation between the two. As Mr. Woodward suggested, cameras in police cars can have a salutary effect on the relationship between police and the public.

It seems to me cameras that record certain events can clear innocent people, when witnesses testify to things that aren't accurate and the cameras demonstrate it. Better evidene leads to pleas and gets rid of cases fairly early if there is documented evidence of the criminal activity that might otherwise be fought.

So it may be possible, for example, not to have any conflict where there is an indication that cameras are up, but in fact they are not working, they act purely as a deterrent, without any compromise of personal privacy. So, you know, I think there is room for a wide range variation.

Ms. NORTON. The variation you suggest suggests that if somebody doesn't put some limits on what those variations may be that they will be all over the map.

Mr. GOLDSTOCK. That is right. I think the limits should be there. I think that is what Congress should be doing. It is minimum standards and limits. You can operate between certain guidelines, and then you have to make determinations based on your own particular needs just where you are going to be.

Ms. NORTON. Mr. Barnes.

Mr. BARNES. Of course, Congresswoman, the Mayor and D.C. council have the authority to set standards and limits with respect to the Metropolitan Police Department. But we have also—we also know that the Secret Service is there, the Park Service, U.S. Marshals Service. The Park Service, you know, Mr. Parsons testified. So there are areas that are purely within the purview of the Congress in terms of legislating and providing some guidance to the Federal agencies. But we would of course prefer the local government to provide that guidance to the Metropolitan Police Department.

Ms. NORTON. At the very least, I think the District should proceed, and perhaps they can—what they do can be instructive to anything that the Chair and I might agree upon.

Look, the District has the largest police force per capita in the United States; that is even if you don't count the Federal police here, has always had a very high crime rate. As I speak, you know, there has been a spike in crime in one neighborhood, and people are calling for more cops.

One wonders—I am back to my example—that there is a certain point at which 10 cops we have no objection to. But one camera that does what 10 cops would do we do. I hate to put another law professor hypothetical to you, but at some point one has to come to grips with the fact that at least in some of the instances we are talking about; for example; Mr. Parsons gave a very hard case, because there is no question that you could have a park policeman right there inside the Washington Monument, they are looking for somebody who would go and put some kind of explosive device there, and one way to do it is to station yourself a police officer there.

I will give you another example. I am in discussions with the House on their closing the stair—West Front stair, hideous thing to do. This is one of the great history vistas. It really goes back to L'Enfant. I am in discussions about opening it up at least some of the time, and we are getting somewhere on how to do it.

Now, what I have suggested is you could put a police officer—it would be a Capitol police officer there for the period in which it would be open, and then we could open it at least for that period. Now, what am I to do if the Sergeant at Arms, with whom I have had the most serious discussions, says, well, Congresswoman, we have just got enough police to cover—it took us a long time, and only September 11th got us enough police to cover what we need.

But I could have a camera there and the West Front could be open the way it always was. So I put that hypothetical to you and ask you if that is the suggestion that I am offered, should I take it, given the outcry that one of the great vistas to the Capitol and to our city has been closed because there is a real danger. You really could come up these stairs and put a device right under a major part of what the Capitol is just by walking up the stairs.

What are you going to do with that one?

Mr. BARNES. Well, again, Congresswoman, I think that is why it is important to really study whether or not these cameras make a difference or do they give us a false sense of security. I know, for example, they had for over a 22-month period cameras were trained in Times Square, concentrated cameras. Ten arrests for petty crimes. I know of no evidence where these cameras have ferreted out terrorists. No evidence where those cameras have really been effective in preventing, deterring, helping with major crimes.

And so the question is, how do we spend our money? And I think we can look to Detroit, for example, where after a decade and a half decided to spend their money in other ways. We can look to Oakland that studied at great length this issue. And the Chief of Police finally said, it is not worth it. We can look to Newark, White Plains, many places throughout the United States. Even Tampa has abandoned, at least for the moment, the facial recognition program that it had 2 years ago during the Super Bowl.

So these adventures have been tried and abandoned, and we need to know whether they are cost effective. And particularly when you measure the cost, dollars and cents—not only the cost in terms of dollars and cents, but the costs to our privacy. The burdens far outweigh the benefits in our opinion.

Ms. NORTON. I think you raise a good point. And this will be my last statement about costs. Of course the cost of stationing a cop is there, a very substantial cost as well, and having him look at everybody who goes up and down is the kind of, "invasion that I did not have when I walked up the Capitol steps before."

I am going to ask the Chair, because I think Mr. Barnes has raised an important question about effectiveness that we have not had answered here. It is true that the Chief said that 10 percent of those arrested in Sydney, which is a very high number, 10 percent of the arrests came from this camera. And he mentioned things like assaults and the rest of it.

I would like to ask the Chair if we might write to Detroit, to the locations that have been named so that we might add to the record why they abandoned this. For all we know the technology might not have been as effective then. But we need to know why it was abandoned, because the whole notion of effectiveness seems to me to be at the root of this. If we are just putting some cameras up there to make ourselves feel good, than it does seem to me we ought to take them down right away.

Thank you very much, Ms. Chairman.

Mrs. MORELLA. I think that is a good idea, Ms. Norton. We may want to, without overtaxing the GAO, ask them to look at how effective these surveillance devices are.

And I want to thank this panel for the expertise you have given us. I know there are a number of other questions we would like to get to you.

As an aside before I just conclude, Mr. Barnes, what is the ACLU's position on the use of red light and speed cameras if they are tied into a camera surveillance system like the District's? Do you have a position on that?

Mr. BARNES. Well, no, we haven't taken a position. But I will say this, Madam Chair. The difference between the red light cameras is that they focus on a specific individual who allegedly has run a red light in a particular circumstance.

That is very different from the general focus on all of us that this video camera surveillance system would do.

Mrs. MORELLA. Very good. We will be calling on you if we do draft any legislation to give us your expert response and advice as we move along, if that would be acceptable with you. We feel that we have some good minds here with adequate background, experience, and that you could enhance what we may want to do.

So I do want to thank you all for being here. I am going to adjourn this subcommittee. I think it has been an excellent hearing. I want to recommend accolades for our staffs. On the majority side Russell Smith, the Staff Director. Robert White, Shalley Kim, Matthew Batt, Heea Vazirani-Fales.

On the minority side, John Bouker and Earley Green. Thank you all very much.

[Whereupon, at 11:40 a.m., the subcommittee was adjourned.]

[Additional information submitted for the hearing record follows:]

SIA
SECURITY INDUSTRY

April 18, 2002

The Honorable Constance A. Morella (R-MD)
Chairwoman
Subcommittee on the District of Columbia
B-349C Rayburn House Office Building
Washington, DC 20515

Dear Chairwoman Morella:

On behalf of the Security Industry Association (SIA), which represents over 400 electronic security equipment manufacturers, distributors, and service provider organizations around the country and throughout the world, we are submitting these written comments for the permanent record regarding the March 22, 2002 public hearing conducted by the subcommittee on the "D.C. Surveillance Network." We appreciate your including SIA's comments and observations on this most important issue.

There is perhaps no more effective security tool than closed circuit television (CCTV). But with the use of CCTV, both in private, corporate use and in public or municipal use, comes responsibility. To this point, the Security Industry Association (SIA) and the International Association of Chiefs of Police (ICAP) teamed to write guidelines for the proper use of CCTV equipment.

Several United States and European public safety models have demonstrated that closed circuit television (CCTV) is a critical component of a comprehensive public safety and security plan. Although, in the U.S., the constitutionality of CCTV use in public areas is well established, there are nonetheless concerns within the public arena with regards to the implications of CCTV use on privacy and civil rights. To consider these issues and develop a guideline regarding the appropriate use of CCTV technology within the public sector of the local community, IACP/PSLC conceived a CCTV for public safety summit. The Summit involved CCTV manufacturers, law enforcement organizations, civil liberty organizations, tort and constitutional lawyers, state and federal regulators, state and federal legislators, and local citizens groups. The commentary from the summit participants was instrumental in creating the final version of the guidelines, attached here as Appendix A.

The proposed guideline advocates that each law enforcement agency coordinates the intended purpose and focus of its CCTV program with the community in which the CCTV program will reside. This exchange of information and knowledge should clearly define and outline what problems the CCTV program was designed to address; it should become the guiding policy document. This policy should then become part of a public document

Security
Industry
Association
635 Slaters Lane, Suite 110
Alexandria, Virginia
22314-1177
Telephone: 703.683.2075
Facsimile: 703.683.2469
www.siaonline.org
www.SecurityGateway.com

distributed to the community. It is recommended that these programs subscribe to existing laws and accepted procedures of evidentiary gathering.

Expressed Purposes for the Use of CCTV

It is vitally important for a jurisdiction to clearly articulate the specific purpose for the implementation of CCTV prior to its use. In so doing, the jurisdiction will be able to identify the best camera, lens and lighting necessary for their specific purpose. In some rare cases, it may be that the designated purpose does not lend itself to the use of CCTV at all. In those cases, the taxpayers of the jurisdiction would be saved a great deal of money and perhaps a lot of needless public debate. In all other cases however, it is essential to know what needs to be accomplished by the technology and to match the specific technology with the desired goal of its implementation.

Generally speaking the purposes of the CCTV technology can be grouped in several different ways. One way of grouping the technology would be to identify the specific purpose of the installation as opposed to who uses the results. For example, to curb vandalism, to identify traffic violators, to identify thefts and other property offenses, to reduce violent crime and to help stop terrorism. These represent some of the desired goals of CCTV and it is from this typology that the following discussion of the current utilization of CCTV Guidelines (see Appendix A) and researched will follow.

Public Safety

Some jurisdictions identify public safety as the reason for the use of CCTV. The University of Pennsylvania (Penn) employs CCTV as a meansof insuring the safety of students on its campus. In a rather unique application, Penn has tied cameras to some of its 66 cellular phones placed around the campus for the safety of its students. When a phone is picked up from its holder, a camera will automatically swing to get an image of the phone and the caller to identify the nature of the problem. In the near future, Penn hopes to have all 66 cellular phones in sight of cameras so that an immediate response to calls for assistance can be assured.

Another example of the public safety rational for the use of CCTV is in Runnymede, England. The jurisdiction of Runnymede is outside of London and is somewhat rural. The CCTV cameras are used in public areas of various cities in this jurisdiction but the real public safety issue is that the control room also responds to "panic buttons" used by elderly citizens. For a very nominal fee (or the fee can be waived for the poor) an elderly individual is given a device to wear that will activate contact with the control room in case of an emergency. When that emergency is a fire or an intruder, the cameras can swing to where the person lives and get a picture of the outside of the house/apartment and send help. Likewise, Newham, England's CCTV control room handles the equivalent of 911 calls during the evening hours. In Newham the typical 911 call during the day would be handled by the city government, but during the night it falls to the CCTV operators to handle emergency calls.

Other examples in the public safety area include crowd control. This purpose was cited in Ealing, England as well as Philadelphia (Penn). In the case of Ealing, soccer matches are often cases for the use of the CCTV technology for crowd control. It is often reported that the

British are enthusiastic soccer fans and that enthusiasm frequently gets out of hand. Ealing has a "war room" where all cameras, communication and deployment of law enforcement are centrally located and while the "war room" may also be used in cases of terrorist activity, the more frequent application is in times of rowdy soccer fans. Likewise, crowd control is a purpose cited by the University of Pennsylvania for its use of CCTV. Penn has about 14,000 students living on campus and cameras assist in the surveillance of public areas where large numbers of students congregate.

Property Crime and Vandalism

Frequently cited as a reason for the use of CCTV is the ability of the technology to curb the rate of property crimes and vandalism. In Runnymede, England a specific effort was made to decrease the amount of property crimes in public areas with the use of cameras. Likewise, in Woking, the central shopping district was targeted in an effort to reduce the amount of property crimes facing shoppers. In Tampa, Florida the use of CCTV in an entertainment area was designed to reduce the effects of con artists and pickpockets upon patrons to the area. At Penn in Philadelphia, vandalism of property was a reason cited for the use of CCTV. Certainly any jurisdiction worried about the proliferation of vandalism could easily measure its effectiveness by comparing the amount of money spent to repair the effects of vandalism before and after the implementation of the cameras. Interestingly enough, no jurisdiction to my knowledge has ever done that simple task.

General Detection of Crime

Most jurisdictions have identified the detection of crime in general as a reason for the use of CCTV. New York City, for example, has chosen to place their cameras in public housing developments as one tool to reduce the amount of crime. They claim a significant decrease in the number of serious index crimes occurring within these developments and that citizens feel safer as a result. Likewise, Jersey City, New Jersey has also chosen to place cameras in areas that had a history of high crime.

Terrorism

In the wake of September 11th, the United States is beginning to deal with issues of terrorism that have plagued other parts of the world for decades. As a result it is not surprising that terrorism has been a stated goal of CCTV in many jurisdictions in England. In Ealing, the site of the most recent terrorist bombing in England, cameras actually caught the event happening. While cameras may not effectively stop the event from happening, they will assist those investigating the crime by allowing for identification of the terrorist and perhaps those that gave assistance to him.

Miscellaneous

One unique reason identified for the use of CCTV is in Tampa, Florida. Because CCTV cameras are used in an entertainment district, and some of those cameras are equipped with face recognition capabilities, officials cite the identification of runaway juveniles as a reason for the technology. To date, this is the only jurisdiction that has referenced this reason for the deployment of CCTV.

Probably the most frequently articulated reason for the use of CCTV is the ability to enhance law enforcement capabilities. I have alluded to that reason above, but specifically, the cameras become the extra eyes of the police. They allow for officers to be deployed in other areas where crime may be better discovered by traditional police work and save for the cameras and their operators the more mundane tasks of traffic and general surveillance. In situations such as this it is a win-win for all involved.

Perhaps of all the cities visited or called to date the clearest goals and statement of purposes comes from the University of Pennsylvania in Philadelphia. They clearly state that the reason for their use of CCTV is to assist law enforcement officers, ensure student safety, to facilitate car stops, observe suspicious activity, crowd control, officer safety and to verify alarms. Clearly, Penn has thought through their technology and the reasons for its use in a fashion that other jurisdictions could well learn from. As will be seen below, by clearly articulating the reasons for the technology, the proper cameras, lens, lighting and placement money is saved and the overall success of the venture is enhanced.

Type of Equipment Currently Used

There are a variety of options available on the market from which to choose. Most areas visited use either analog or digital technology with color cameras. Several jurisdictions (Tampa, Florida for example) still use some black and white cameras, but those are being replaced as funds become available.

In most areas there is a combination of fixed and pan/tilt/zoom cameras. The fixed cameras are most often found in parking garages or to identify red light runners. The pan/tilt/zoom cameras are most effective for surveillance of public areas by allowing the operator to follow a suspicious person and zoom in on them to identify the activity being observed.

In areas that use digital technology there is a wide difference in the number of frames that are captured per second. For example in Woking, England the digital technology is such that only 4 frames per second is recorded. As a result, the finished product looks "choppy" and you do not get a clear view of an entire event. A better result can be found in Philadelphia at Penn where they record 15 images per second with their digital technology. At 15 images per second some of the "choppy" results seen in England are avoided. However, the best result can be found in New York City where they record 30 images per second and this certainly is the closest to real time recording of any observed.

Several jurisdictions use mobile cameras in addition to the fixed and the pan/tilt/zoom models. The mobile cameras are primarily used for covert operations but can be moved to "high crime" areas if the need arises. In jurisdictions that utilize mapping technology such as Ealing and Penn, cameras that are mobile can be used as an immediate response to the results of current crime statistics. Thus, by having cameras that are not stationary and by using crime-mapping software, a jurisdiction can respond quickly to changes in crime patterns.

Finally, some jurisdictions have built what they refer to as "war rooms", usually as an adjunct to their control rooms. In some rare instances these may be cost effective if used frequently. The "war rooms" are equipped to allow observations of the camera's monitors and to coordinate responses from a central location. They are most often hooked up to law enforcement, fire and emergency agencies and thus allow for communication between these various agencies in times of civil strife or major events. "War rooms" were observed in Ealing and Newham, England and they were not used with much frequency.

Who is Responsible for Monitoring?

Generally, the monitoring of the CCTV equipment falls into three classifications: law enforcement, private companies or civilian employees of local government. Each structure has advantages and some disadvantages and it is up to the local jurisdiction to identify the best model for their own purposes.

Law Enforcement

Law enforcement was responsible for the monitoring of CCTV equipment in three of the jurisdictions visited. In Woking, England law enforcement monitored the cameras in a control room that was physically separate from the police station itself. Only one officer was responsible for the operation and monitoring of cameras. In Tampa, Florida law enforcement was also responsible for the cameras. In New York City law enforcement officers who had been hurt on the job or who had disciplinary issues yet to be resolved were the ones monitoring the cameras. With one exception (a housing development in the Bronx) all the housing developments used law enforcement officers for monitoring the equipment and have the control room for the cameras on site at the housing development. Thus, police officers actually work in the housing development that they are charged with observing.

While there are advantages to having law enforcement charged with the observation of the monitors, there may be some problems that this arrangement fosters. For example, the response to potential crime may be quicker with law enforcement doing the monitoring since it may decrease the amount of time between observation and deployment of officers. However, law enforcement officers may be more sensitive to real "crime" issues as opposed to responding to citizens in need of assistance. Another issue related to law enforcement officers manning the monitors is that in all cases SIA is familiar with, the training was on-the-job. Thus, all information that these operators have is from a law enforcement perspective and that may hinder or restrict the utility of the technology.

Private Companies

In Ealing, England a private company was used to do all monitoring of the equipment for general surveillance and for traffic control. The employees of this company were trained by that company and then had a short period of on-the-job training to meet the needs of the individual jurisdiction that utilized their services. Likewise in Philadelphia at the University of Pennsylvania a private company was used for monitoring purposes as well as the up-keep of the equipment. The value of such a scheme is that the monitors are not "entrenched" in a law enforcement mentality and appeared to be very willing to see their job as a cross between law enforcement and social working for the public. As a result, their goal is not merely to assist in making arrests but also to provide a service to the people in the area where they are assigned.

Civilian Employees of Local Government

In Runnymede, England employees were selected for these positions based upon their desire to perform services for the public. For example, as identified earlier, the employees are responsible for dealing with the "panic buttons" that elderly individuals are given in addition to the regular duties of surveillance. These employees must demonstrate good social skills because they will be dealing with not only police but with the fire departments, emergency equipment supervisors as well as the public on a daily basis. In Newham, England civilian employees work for the city council. These employees are not only trained on the use of CCTV technology but with the use of face recognition software as well. These employees did not have the emphasis upon the social work function of their position to the degree found in Runnymede and that is probably due to the fact that Newham is a much more urban environment than Runnymede. Finally, in Jersey City civilian employees are used to monitor the technology and they work in a control room that is not located within the local law enforcement building. That arrangement was by design as the goal was not to give the population the impression that the police were watching them.

The general advantages of having a group of civilian employees responsible for monitoring the technology is that they are not police officers. This may make the use of CCTV, especially in the United States, more acceptable to the public. Additionally the training is often supplied by a company that specializes in training for all kinds of CCTV purposes and that will allow the person going through the training to get a perspective that is not totally law enforcement. There are some disadvantages to the use of civilian employees. For example, there may be reluctance on the part of law enforcement to see these employees as assisting in the policing efforts. If this occurs there may be negative consequences for the citizens due to communication problems between those that are monitoring the equipment and those that must respond to calls for assistance.

Rules Governing the Technology

There are differences in the length of time that tapes are kept for potential criminal proceedings and the number of times that tapes can be reused. Since England has been using CCTV technology for well over a decade it is not surprising that they have developed more specific rules that govern the issues of length of time tapes are kept and reuse patterns. Most jurisdictions keep their tapes for either 28 or 30 days. This coincides with the mandated rules set down by the CCTV Users Group and the government. There are also specific guidelines on the number of times that tapes can be reused. In all cases tapes are reused twelve times and no more.

In the United States there is much more variation in the length of time tapes are kept and the number of times those tapes can be reused. For example, in Philadelphia at the University of Pennsylvania, tapes are kept for 15 days but they are reused 50 times. In New York City on the other hand, tapes are kept for 7 days and are reused only 7 times. This goes to show how badly the United States is in need of clear guidelines that will result in the same treatment of tapes for storage and reuse purposes.

Each agency has and must have clear protocols for the chain of custody of tapes that will be used in criminal proceedings. While they may vary slightly from one jurisdiction to another, they all have specific rules and those rules reflect the standards that are set locally by the government and the courts.

Training

The type and degree of training varies greatly between agencies. Most agencies employ some type of on-the-job training. Even when the employees receive training from an outside company there is a period of training that takes place once the employees reach the site of their monitoring. That makes sense since each jurisdiction has certain oddities that are unique to their area. Some of the on-the-job training lasts just a day or two (New York City) while others involved at least one week (Runnymede).

Private companies have been formed in England to supply the training for local jurisdictions. They may train their own employees to be assigned to jurisdictions or they may train civilian employees for local government. The only area in England that does not use the private companies was in Woking where law enforcement did the monitoring and trained their own employees on-the-job.

A frequent strategy employed in areas is to get training from a company or through the agency that hired the monitor and have that training followed by some amount of on-the-job training. Most agencies indicate that when they would get new cameras or change the database, more training would follow to update their skills.

Data on Effectiveness

Sadly very few jurisdictions are keeping any kind of data to demonstrate their CCTV technology is effective. Currently, SIA has engaged a researcher to study the effectiveness of CCTV systems in the United States and England. This extended survey will compile the kind of scientific data necessary to truly gauge the effectiveness of CCTV. The general consensus is that CCTV is effective, but most of that information is anecdotal. Likewise there is a general belief that the use of CCTV is a deterrent to crime but again, no statistics seem to verify this. There are also very common claims that the number of arrests have increased since the deployment of

CCTV but again, little or no statistics are used to back this assertion. The information that follows must be judged based upon the quality of the data. Some of this information is questionable and some is more scientific in nature, although none of the jurisdictions have employed the true scientific method in the compiling and presentation of statistics.

In England, even with the longer history of the use of CCTV, most jurisdictions are not keeping statistics on the amount of crime that was discovered as a result of CCTV or on the number of arrests that were the direct result of CCTV. In Woking, England where law enforcement runs and monitors the technology no statistics are kept at all. They suggest that crime has decreased as a result of their efforts, yet there are not clear data gathering methods to back up this assertion. In Ealing, England there are some general crime statistics kept and they do use mapping software to allow for the moving of cameras to respond to crime but the techniques used are far from scientific. For example, there is no way to decide how much of the crime in their area is discovered through the use of CCTV since all crime is lumped together and there is no way of deciding the proportion of crime discovered through CCTV, through law enforcement or through citizens. In Newham, England which uses CCTV and face recognition software there are general crime statistics kept. In the four years since face recognition software has been used only one person has been identified and caught as a result of this technology. Overall they claim to be very successful in the recovery of stolen vehicles and tags and they also indicate they have had a 34% reduction in crime as demonstrated with over 20,000 fewer offenses recorded.

One exception to the lack of statistical data is Runnymede, England. Monthly reports are generated of all CCTV activity that includes the area of the discovery, the date, the time of day and the description of the behavior observed. They even attempt to find out the disposition of the individual observed from law enforcement once the event is turned over for their handling. Their data is broken down by the events that are law enforcement discovered, discovered through a third party or those that are discovered by CCTV. Their overall findings are that within a three year period, CCTV has resulted in about 7,000 offenses being discovered leading to nearly 1,000 arrests. They report offenses by type of crime within each jurisdiction and also the number of arrests per type of crime within each jurisdiction. This data is the result of an initiative by David Dodd, Manager of the Technical Services Department, and is not required by the government. As such, his is one of the few efforts being made to attempt to measure the effectiveness of this technology.

In the United States law enforcement departments supply crime data on the number of crimes known within their jurisdiction to the federal government for classification based on type of crime. Since the data is sensitive to political issues among other things, the academic community has long been suspicious of this data. With that being said, it is still significant to view this data since it is considered the only "official" measure of crime in this country. Over the years law enforcement departments have structured their crime gathering and reporting information in the form required by the federal government. As a result it does not lend itself to measurement for the purposes of identifying whether or not CCTV is effective. In other words, crime data is lumped together with no thought given to the origin of the discovery of an incident. These issues must be kept in mind when discussing the crime statistics available in America.

In New York City law enforcement officials claim that within the public housing facilities where CCTV is being used, 1,200 arrests have resulted in four years. Likewise they claim a significant decrease in the number of index crimes occurring within nearly every housing development where CCTV is being used. The rates of change given to me from the New York City Police Department range from a modest increase of 14% in the Stapleton Development to a decrease of 50% at Carey Gardens. Overall they claim a system wide decrease of 24% in index crimes within the housing developments serviced by CCTV.

In Jersey City, New Jersey 450 incidents were discovered by CCTV that required police assistance within a six-month period ending in January 2002. As a result of the discovery of these 450 incidents, 150 arrests were made. There was a shooting, numerous drug deals and robberies that were discovered through CCTV and ended with arrests.

The facility at the University of Pennsylvania in Philadelphia has come the closest to measuring the incidents discovered through CCTV of any jurisdiction within the United States. Overall, they claim a 4% reduction in crime, about the same as the city of Philadelphia as a whole. Where they believe the most important information resides is in the increase in arrest rates that have resulted since the deployment of CCTV. The Penn facility, unlike others, uses crime mapping software and a CAD program that allows them to identify the source of the incident. They currently have the capabilities to separate out the incidents discovered by CCTV, by law enforcement and by third parties. Most of the CCTV crime that was discovered occurred during the 7 to 11 shift. They have noticed a reduction in vandalism as well as in armed robbery. Armed robberies have decreased from a high in 1996 of 129 incidents to a low in 2001 of 4 incidents. They still have work to do on the accurate reporting of crime that can be discovered through CCTV but they appear to have much of the technology in place to accomplish this endeavor.

Overall, it is fair to say that no jurisdiction is currently keeping the kind of statistical data that can be analyzed in such a way to demonstrate the effect of CCTV. There are certainly software programs on the market that can assist with this effort and some jurisdictions may also have the in-house capabilities of developing programs on their own to measure its effect. Part of the problem seems to be that the software currently on the market may not work well with the database that is being used to send crime data to the federal government. This topic will be the focus of future study.

This technology, when applied carefully and properly, and governed by appropriate regulations, standards, and guidelines, will assist municipalities in reducing crime and criminal behavior. This technology will also assist our local governments in attempting to prevent future terrorist events from occurring

Sincerely,

Richard Chace
Executive Director, SIA

Appendix A

Overview on the Guidelines for Closed Circuit Television

An Overview on the Guidelines for Closed Circuit Television (CCTV) For Public Safety and Community Policing

Promulgating the Responsible Use of CCTV Technology in Security and Public Safety Applications

Prepared by
Richard W. Chace
Executive Director
Security Industry Association
Alexandria, VA
E-mail: chace@siaonline.org

Table of Contents

Forward 3

Overview of Guideline Development 5

CCTV for Public Safety and Community Policing Guideline 10

Building Productive Public-Private Partnerships 18

Supplemental Legal Information 21

Sample Case Study 47

Resources 49

Forward

The CCTV for Public Safety and Community Policing Guideline has taken three years to complete and would not have been possible without the commitment and support of the International Association of Chiefs of Police's (IACP) leadership, staff and Private Sector Liaison Committee. The IACP has been outstanding and uncompromising in their quest to positively effect how technology can be most appropriately and responsibly used in furthering the causes of law enforcement. The Security Industry Association (SIA) feels fortunate to have made such an historic partnership with this prestigious organization, which responsibly addresses the issues associated with the use of closed circuit television (CCTV).

The achievements of the Private Sector Liaison Committee (PSLC), under the steady leadership and direction of Chief Michael Shanahan, have been truly outstanding and beyond reproach. The members of the PSLC unselfishly give of their valuable time and knowledge to work on programs designed to improve law enforcement's interaction with the private sector. In similar character, the members of the PSLC's CCTV Sub-committee, under the strong and careful leadership of Chief Steven R. Harris (also a former president of IACP), have rallied to draft and shepherd the many revisions of the CCTV for Public Safety and Community Policing Guideline to completion and widespread dissemination.

Specifically, the guideline would not be in existence today if it were not for the dedication and expertise of the following people: Chief Michael Shanahan (Ret'd); Chief Steven R. Harris, Redmond, Washington; Thomas M. Seamon, CPP former Vice President of Public Safety University of Pennsylvania and the CCTV guideline's primary architect; Robert Bickel, Professor of Law at Stetson University and the CCTV Sub-committee's primary legal advisor and author of the CCTV Legal Memo; Richard Chace, Executive Director of the Security Industry Association and coordinator of the 1999 CCTV Summit and author of *CCTV for Public Safety* compendium; Richard Moe; Vice President of the Interpro Group; Ira Sommerson, President, Loss Management

Consultants; Marene Allison, Vice President Loss Prevention at Great Atlantic & Pacific Tea Co.; Chief Joseph Dunne, New York City, New York; Chief Michael Brassfield, Fort Lauderdale, Florida; James Harris, President, Regent International Solutions; Ronald Schwartz, CEO Universal Atlantic Systems; Ronald Spiller, Executive Director, Security Industry Association; Lessing Gold, Esquire, Mitchell, Silberberg & Knupp, LLP; Jeffrey Blum, Vice President of Strategic Planning, Ultrak; Chief Bruce Glasscock, Plano, Texas; Daniel Rosenblatt, Executive Director, International Association of Chiefs of Police; Eugene R. Cromartie, Deputy Executive Director, International Association of Chiefs of Police; Charlie Higginbotham, Director of Information and Services Division, International Association of Chiefs of Police; John Firman, Research Coordinator, International Association of Chiefs of Police; Jeffrey Higginbotham, Chief Legal Counsel, International Association of Chiefs of Police; and the many participants in the 1999 CCTV for Public Safety and Community Policing Summit.

Overview of Guideline Development

The manufacturers and distributors of closed circuit television (CCTV) security products, represented by the Security Industry Association (SIA), and members of the law enforcement and public safety communities, represented by the International Association of Chiefs of Police (IACP) and the National Sheriffs Association (NSA), are committed to enhancing the quality of life of the local community by integrating the best practices of public and private policing with state-of-the-art security technology.

Several United States and European public safety models have demonstrated that closed circuit television (CCTV) is a critical component of a comprehensive public safety and security plan. Although, in the U.S., the constitutionality of CCTV use in public areas is well established, there are nonetheless concerns within the public arena with regards to the implications of CCTV use on privacy and civil rights. To consider these issues and develop a guideline regarding the appropriate use of CCTV technology within the public sector of the local community, SIA and the IACP Private Sector Liaison Committee conceived a CCTV Summit. The Summit involved CCTV manufacturers, law enforcement organizations, civil liberty organizations, tort and constitutional lawyers, state and federal regulators, state and federal legislators, and local citizens groups.

At present (January 2001), there are an estimated 2 million + video cameras in use around the country for the purpose of promoting public safety and security. Many of these cameras have been in use for years in applications such as Automatic Teller Machines (ATM's) and traffic regulation. Despite the prevalence of CCTV use on the national and local levels, there have been (prior to the Spring 1999 CCTV Summit) no consistent policies or procedures guiding the use of this equipment. Given the ethical, legal and other important issues implicated in the use of CCTV technology in the public sector, the members of SIA, IACP, and NSA recommend that public safety officials and law enforcement agencies adopt some or all of the following written guideline to assist and facilitate in the use of CCTV technology within the local community.

Background of Closed Circuit Television (CCTV) Use

CCTV has been widely used in public areas by law enforcement and private security organizations in the United States. Currently, CCTV technology is being used by city police departments, such as New York and Baltimore, and on University campuses, such as the University of Maryland at College Park and the University of Pennsylvania. Much of the existing CCTV use at the local level is currently being used to monitor traffic; especially traffic signal-controlled intersections and to observe and sanction aggressive driving.

Critics of CCTV uses in the public sector have raised two constitutional issues: 1.) the Fourth Amendment guarantee against unreasonable searches and seizures, and 2.) the right of personal privacy, a generic term encompassing various rights recognized to be inherent in the concept of ordered liberty under the Fourteenth Amendment. The clearly established constitutionality of CCTV use in public areas rests on the concepts of "public area" and "reasonable expectation of privacy," as defined extensively in case law. Generally, public areas are those areas open for public use, including unenclosed areas (public streets, sidewalks, and parks, etc.) and enclosed areas (building lobbies, corridors and elevators, etc.) To qualify as a constitutionally protected "reasonable expectation of privacy," the individual must have an actual expectation of privacy and that expectation must be one which society recognizes as reasonable.

The courts have consistently found that an individual does not have a reasonable expectation of privacy when he or she is in a public place. Behavior and activity exhibited in a public area is obviously available for observation by others. Police observation of activities conducted in plain view in a public place, therefore, does not violate the Fourth Amendment guarantee against unreasonable search and seizure, regardless of whether the observation occurs through the physical presence of a person at the scene or through the assistance of CCTV technology. Similarly, there is no violation of personal privacy rights under the Fourteenth Amendment when an individual's public behavior is observed by a video camera.

However, it is important to re-iterate, regardless of the green-light given by current law, responsible and ethical use CCTV technology as a public safety and security tool is critical to the success of current and future public safety applications of CCTV and other technologies. SIA, IACP and the NSA are firmly committed to promoting such use and strongly urge all law enforcement agencies actively using or contemplating the use of CCTV technology to use the CCTV for Public Safety and Community Policing guideline.

Common Questions and Answers

1.) **Will there be security cameras in public bathrooms or other areas designated as "public," where an individual may expect privacy?**

No. Despite the name "public restroom" or "public bathroom," the proposed guideline recognizes these spaces in which one has a reasonable expectation of privacy. The proposed guideline prohibits CCTV use in areas where there is a "reasonable" expectation of privacy, as defined by existing law. This guideline would, however, permit CCTV use in the hallway or area outside a public restroom or similar use facility.

2.) **How will CCTV use change the way law enforcement patrols and interacts with my community; will such use supersede law enforcement's current means of street patrol?**

The proposed guideline advocates that each law enforcement agency coordinates the intended purpose and focus of its CCTV program with the community in which the CCTV program will reside. This exchange of information and knowledge should clearly define and outline what problems the CCTV program was designed to address. This should then become part of a public document distributed to the community. It is recommended that these programs subscribe to existing laws and accepted procedures of evidentiary gathering.

However, if an individual(s) is/are perpetrating a crime in a public area they may still be stopped or addressed by law enforcement, but not necessarily due to the operation of CCTV equipment. The use of CCTV technology in public areas is intended to be a force-multiplier designed to assist law enforcement in the execution of their duties.

3.) **How do I know law enforcement is not using CCTV technology to track my normal daily activities and movements?**

The proposed guideline places a great deal of emphasis on individual privacy and rights. Subsequently, the vehicles/tools used to store image data are subject to specific handling protocols. In order to gain public support of CCTV use, law enforcement agencies should adhere to a specific operational guideline, specifically the CCTV for Public Safety and Community Policing guideline. This guideline, among other specifics, clearly states that normal CCTV-obtained images should be purged on a regular basis and retained in accordance with applicable public record laws. This affords law enforcement a fail-safe in case an image obtained through CCTV technology becomes a piece of evidence. Law enforcement, as a general rule, can only use its time and resources to identify instances that require action based upon "just cause."

4.) **Who wrote this draft guideline?**

The proposed guideline has been developed over the last two years through numerous discussions and meetings by the International Association of Police Chief's (IACP) Private Sector Liaison Committee whose members include representatives of the law enforcement, public/private security professional, CCTV manufacturer, legal and regulatory communities. This committee considered the ethical, social, legal and practical implications of CCTV use for safety and security purposes. A draft guideline was constructed based on the committee's discussions, the University of Pennsylvania's CCTV Monitoring and Recording of Public Areas for Safety and Security draft policy, and the United Kingdom's Metropolitan Police Service Public Place CCTV Systems Guidance Guideline.

The resulting draft was prepared for discussion and debate at a CCTV Summit in the Spring of 1999 in Washington, DC. There, over a two-day period, members of the law enforcement, public/private security professional, CCTV manufacturer, legal and regulatory communities edited the guideline and offered revision suggestions to key elements of the document. The IACP's PSLC CCTV Sub-committee reviewed the recommended edits over a two-day period, incorporating suggested edits into the draft guideline document. The resulting 5th Revision was presented to the PSLC with the recommendation that the draft be moved into the three-month Review and Comment Period, starting in June 1999 and ending in September 1999. The responses from this three-month comment period were then reviewed by the PSLC CCTV Sub-committee and incorporated as necessary. The resulting 6th revision was then reviewed for continuity and substance. The final 7th revision was presented to the full PSLC with recommendation that it be approved and disseminated to all interested constituencies.

5.) What was the CCTV Summit and who participated?

The two-day CCTV Summit, held April 8-9, 1999, Washington, DC, at the Capital Hilton, hosted representatives of the CCTV manufacturer, public/private security, law enforcement, legislative, regulatory, civil-libertarian, and legal communities who came together to develop a consensus CCTV for Public Safety Operational Guideline document.

The goal was to create a document that discusses the privacy and legal issues associated with the public safety and community policing applications of CCTV technology.

Security Industry Association

And

International Association of Chiefs of Police

Number 9

Last Revision Date: 1/1/00

GUIDELINE: Closed Circuit Television (CCTV) for Public Safety and Community Policing

PURPOSE: The purpose of this document is to provide guidance to law enforcement in the responsible use of <u>overt</u> closed circuit television (CCTV) cameras in public areas, without a court order, for the purpose of safety and security.

GENERAL PRINCIPLES:

A. In promulgating these guidelines, the security industry and law enforcement agencies seek to establish voluntary parameters restricting the non-court-ordered use of CCTV to public places, to enhance public safety and security in a manner consistent with accepted rights of privacy.

B. Except in situations of the investigation of a crime committed by a person(s) whose description is known, CCTV programs must not be based on individual characteristics, or classifications, including race, gender, sexual orientation, national origin, or disability.

C. These guidelines are intended to demonstrate that the security industry and law enforcement communities are committed to enhancing the public's quality of life by integrating the best practices of public and private policing with the responsible use of technology.

D. The principle objectives of any CCTV program should include:

1.) Enhancing public safety;
2.) Preventing/ deterring crime and public disorder;
3.) Reducing and removing the fear of crime;
4.) Identifying criminal activity;
5.) Identifying suspects;
6.) Gathering evidence;
7.) Documenting police actions to safeguard citizen and police officer rights;
8.) Reducing the cost and impact of crime to the community; and
9.) Improving the allocation and deployment of law enforcement assets.

E. CCTV use for safety and security purposes should be conducted in accord with accepted legal concepts regarding privacy, and in a professional, and ethical manner. Personnel involved in CCTV use should be appropriately trained and closely supervised in the responsible use of this technology. Violations or breaches of any program protocols should result in appropriate discipline and may subject those involved to civil or criminal liability under applicable state and federal laws governing CCTV video monitoring.

F. Initial and ongoing needs assessments should be conducted as a part of any CCTV for safety and security program or protocol. Such needs assessments should consider that CCTV is only one of many tools available in protecting the public's safety-and that other alternatives may be more appropriate or cost effective.

G. Information obtained from CCTV use should be used exclusively for safety and law enforcement purposes.

Information in any form obtained through the use of a CCTV program, or CCTV technology should be handled according to accepted law enforcement procedures and legal rules governing the handling of evidence. Dissemination of such information should be conducted in accordance with applicable State and Federal laws. Unusable or non-case specific video or digital image data should not be retained, and should be purged from data storage within an appropriate time, and in conformance with governing State and federal legal and public policy requirements.

H. Law enforcement agencies should actively seek consultation and input from their community prior to implementing any CCTV program, or any significant expansion or alteration of such a program.

RESPONSIBILITIES

A. Law enforcement agencies implementing a CCTV program shall be responsible to oversee and coordinate the use of CCTV for public safety and security purposes, and shall establish a liaison with their community regarding the program's policies and procedures.

~~B.~~ Each law enforcement agency implementing or using a CCTV program should identify a responsible party for the implementation and oversight of the CCTV program. The designated CCTV oversight officer shall be charged with facilitating input from and conducting consultations with the community. Such consultations should identify the positive aspects of CCTV use, and should work toward securing community support for CCTV use in public places to enhance public safety and security.

C. Any local law enforcement agency implementing a CCTV program should monitor relevant law and security industry

practices to ensure that their CCTV program is consistent with appropriate industry standards and legal protections.

E. Each local law enforcement agency implementing or using CCTV in public places should conduct ongoing program needs assessments and periodic review of CCTV camera locations, perimeter view, monitoring, training, and administration.

D. All local law enforcement agency personnel involved in the application, use or monitoring of CCTV installations, collection of video or digital data, or other aspects of CCTV use shall receive appropriate training, including but not limited to the ethical limits of CCTV use, and instruction in applicable civil and criminal law. Law enforcement agencies and the security industry shall assist in the establishment of standards or criteria for such training programs.

H. All operators and supervisors involved in use of CCTV in public places will be responsible to perform their duties in accordance with applicable law, department or agency policy, and this guideline.

LOCAL LAW ENFORCEMENT IMPLEMENTATION GUIDELINES:

A. All existing public safety and security uses of CCTV technology should be brought into compliance with this guideline.

B. Local law enforcement agencies implementing or conducting a CCTV program shall establish and enforce operating procedures that implement this guideline.

C. Any local law enforcement agency implementing a CCTV program should consider posting signage at appropriate locations notifying citizens that the location may be using

CCTV technology. The posting and content of signage should be reviewed with agency legal counsel.

D. Any use of CCTV to observe locations consisting of residential or commercial housing should limit the view available to that which is only available to the unaided vision of an officer that may be on sight. Furthermore, any view of any residential or commercial housing area must not violate reasonable expectations of privacy, as current case law or statute defines that term or concept.

E. Any monitoring center of a local law enforcement agency implementing a CCTV program must be configured to prevent camera operators from tampering with or duplicating recorded information. Law enforcement agency policies must provide for discipline where this guideline is violated, and must notify all agency personnel that the unauthorized or illegal use, viewing, dissemination, or duplication of video recordings, images, or data, may subject the offending officer to civil or criminal liability.

F. Recorded analog videotape and collected digital video images should be stored for an appropriate time period, consistent with established policy and public records laws, and then should be erased or deleted, unless retained as part of a criminal investigation or civil or criminal court proceedings.

G. Videotapes and digital video images should be stored in a secure location with access, controlled and logged, limited to authorized personnel as defined and designated by the agency.

H. Law enforcement agencies using CCTV must establish and implement programs for the training of personnel involved in the use of CCTV, including camera control operators. Such training shall include technical training related to all equipment and technology used in the program, and shall include all aspects of this guideline.

I. Camera control operators must not use CCTV to track/observe individuals based on characteristics of race, gender, ethnicity, sexual orientation, disability or other classifications protected by law.

TECHNICAL PROCEEDURAL GUIDANCE:

A. In constructing a CCTV program it is necessary to establish each individual program's **Operational Requirements**. These requirements should include:

1.) Identification of areas requiring CCTV use;

2.) Assessment of the number of cameras, their locations and optimum required positions;

3.) Evaluation of existing light levels and positioning of artificial and natural lighting sources in both day and nighttime conditions; and

4.) Choice and identification of the most appropriate camera technology and equipment in relation to the proposed operating environment.

~~B.~~ A system review or audit should be undertaken periodically by accredited and/or qualified personnel, and measured against the specifications developed by each CCTV program's respective **Operational Requirements**. Any such audit must also include an assessment of the CCTV program's compliance with this guideline, including an ongoing assessment of the involvement and support of the community.

C. Any CCTV program must include a system management plan that provides for:

1.) Formulation of control room location, configuration, and staffing;

2.) Development of the CCTV program's functional mission and operational protocols;

3.) Assignment of a supervisor to oversee the operation of control rooms, system equipment, monitors and data collection/storage procedures;

4.) Formulation of security protocols for control rooms, related personnel, equipment and any other component of the CCTV program's system;

5.) In cases of real-time monitoring, formulation of incident response protocols;

6.) Assessment of power supply and backup requirements; and

7.) Formulation and establishment of a routine system maintenance/upgrade program.

POTENTIAL APPLICATIONS OF CCTV IN PUBLIC AREAS:

Legitimate public safety and security purposes may include, but are not limited to, the following:

Purpose	**Example Uses:**
Protection of Persons and Property	Patrol of building perimeters, entrances and exits, public lobbies/corridors/ elevators, public docks, public storage areas;
Monitoring of Access Control Systems	Monitoring of restricted access transactions at

	entrances to public buildings and other public areas;
Verification of Security Alarms	**Confirming, prior to deployment of resources, public building intrusion alarms, trips on exit-door controls, hold-up alarms;**
Video Patrol of Public Areas	**Remotely observe or document activity at transit stops, parking lots, public streets, shopping areas, public parks, school playgrounds, and vehicle intersections;**
Criminal Activity	**Remotely observe or document instances of robbery, burglary, prostitution, vandalism, street crimes, or loitering;**
Traffic Regulation or Control	**Remotely observe and/or document red light running, aggressive driving, and pedestrian/ vehicle traffic at intersections and on major Highways.**

Building Productive Public – Private Security Partnerships

For the past ten years the Security Industry Association (SIA) has been working with the International Association of Chiefs of Police (IACP) as a member of their Private Sector Liaison Committee (PSLC). This sixty member super-committee has been the breeding ground for such industry initiatives as the Model States Program, CCTV for Public Safety and Community Policing Summit, Standards for Mobile Security devices and most recently the Guide for Preventing and Responding to School Violence. In short the PSLC has been a bridge by which the private sector can interact with law enforcement and it's a bridge that has been used to successfully increase the security industry's stature and value within that community.

In the Spring of 1999, in an unprecedented move, the IACP agreed to partner with SIA to co-host a Summit designed to address the issues associated with the use of CCTV in public safety applications. Through prior meetings with key members of the PSLC, SIA was able to deduce that law enforcement was hesitant to institute widespread CCTV programs due to perceived liability and operational issues.

Concomitantly, SIA was aware of outside attempts to limit law enforcement's use of CCTV technology through the courts and certain state legislatures and was concerned that these actions would severely limit this industry's accessibility to a viable CCTV market.

In response to their respective and mutual concerns, SIA and the IACP drafted the CCTV for Public Safety and Community Policing Guideline. This guideline has evolved through 8 revisions, with the 9th approved in early 2000 by the full PSLC and referred to the Board of Directors of the IACP with the recommendation that it be disseminated to police agencies across the country. This guideline has been reviewed by law enforcement, CCTV manufacturers, Department of Justice officials, Congressional leaders, Municipal leaders, Civil Libertarian groups, members of the general public, and Constitutional and tort lawyers. The guideline builds upon a vast reservoir of public safety and legal knowledge and experience and serves as a "How To..." document for those law enforcement and private sector entities seeking to implement a CCTV for Public Safety program.

This three-year effort and comprehensive guideline document has worked to preserve the threatened public safety CCTV market. Perhaps it is not clear to CCTV manufacturers and dealers how dangerously close the CCTV industry has come to losing the public safety market due to the litigation efforts of civil libertarian groups over privacy and security issues. One of the ways to counter such efforts is to promulgate the use the CCTV for Public Safety guideline as an educational and management tool.

The guideline clearly spells out the intent to responsibly promote the use and instruct in the application of CCTV technology while partnering with law enforcement. Additionally, the document recommends ways to institute a CCTV program and how to build one from the ground up.

It is has been vital that the CCTV industry understand it must develop and cultivate working relationships with law enforcement and public safety entities prior to any sales pitches. Six years worth of research, data and personal interaction demonstrating this fact should be respected. Law enforcement agencies are traditionally strapped for funding and information; such agencies are wary of the security industry and the promises made about products and services. The industry may dangle the CCTV carrot in front of public safety officials' noses extolling the force-multiplying virtues of the technologies, but it often fails to outline the need to invest in public outreach and education prior to installing or even purchasing the equipment. Perhaps, even worse the industry over-sells and convinces municipalities and agencies that they need more technology rather than more attention to responsible use of less technology.

Industry data reveals that law enforcement agencies place a higher value on practical information such as to validate or instruct that posting an officer at that intersection during peak traffic hours may solve their red light-running problem.

However, if such resources do not exist, then a less high tech and an inexpensive B/W fixed camera may be the first step in addressing the problem.

Is this a simplistic approach? Of course. But the point is clear: *To responsibly sell to law enforcement at partnership must be established.* Just because manufacturers build a product and sell it, does not mean law enforcement agencies can use it or will buy it. Dealers and manufacturers must demonstrate they are committed to responsibly addressing public safety issues for the long term and can be relied on to provide the right information; even if means offering a lower-tech version of a high-tech product.

The manufacturers and dealers of CCTV technology and equipment can demonstrate their commitment to, and gain trust with law enforcement by participating with groups such as the Virginia Police & Private Security Alliance (VAPPSA) or the Area Police/Private Security Liaison (APPL) in New York. There are such public/private groups in all states whose sole purpose is to develop better working relationships between law enforcement and the security industry.

The Security Industry Association has worked hard to cultivate the trust and respect of the law enforcement community by funding and spearheading such initiatives as the CCTV for Public Safety Guideline program. Now it is up to the manufacturers and dealers of CCTV equipment to avail themselves of this foundational work and provide the services and equipment law enforcement wants, rather than what we think they want.

Supplemental Legal Information on CCTV

Legal Issues Related To Silent Video Surveillance [1]

A Brief Paper on the Subject of
Constitutional Law & Policy Issues, and Tort Liability Issues
Related to the Use of Silent Video Surveillance
To Enhance Policing, and Premises or Employer Security

Prepared by **Robert D. Bickel**[2]
to Facilitate a Discussion of the Subject at a Special Conference
Arranged by
The Security Industry Association and
The Private Sector Liaison Committee
Washington, D.C., April 8, 1999

Introduction to Constitutional Law and Privacy Issues

Legal dialogue among scholars in the fields of constitutional law and the common law of privacy has been ongoing for more than a decade. Early articles on the constitutionality of video surveillance[3] documented the first series of projects, and raised constitutional issues that have been the subject of real outcomes described in the most recent legal commentary.[4] Thus, in ten short years, the legal literature has drawn some fairly solid conclusions, based upon both theory and experience. Similarly, tort law (particularly negligence law) has begun to examine the use of video security systems in the

[1] This memorandum is not intended to provide specific legal advice as to situations in which video surveillance is challenged. Rather, the memorandum is an attempt to summarize selected legal commentary and judicial decisions on the subject.

[2] B.A., Univ. of South Florida, J.D., Florida State University. Professor of Law, Stetson University.

[3] J. Granholm, "Video Surveillance on Public Streets: The Constitutionality of Invisible Citizen Searches," 64 U. Det. L. Rev. 687 (1987).

[4] Q. Burrows. "Scowl Because You're On Candid Camera: Privacy and Video Surveillance," 31 Val. Univ. L. Rev. 1079 (1997).

context of a landowner's duty – as landlord, school, commercial business, etc. – to take reasonable measures to deter criminal activity on the landowner's premises. This outline attempts to summarize the dialogue and identify the most critical legal and policy issues arising from the use of video surveillance.

History

Quentin Burrows notes that video surveillance technology was introduced in certain cities as early as 1956, to assist police in reducing crime on public streets. Early projects included the use of video technology in1966 in Hoboken, N.J., and 1971, in Mt. Vernon, N.Y.[5] Both Borrows and Jennifer Granholm have described these early projects as generally unsuccessful,[6] and Burrows paints a similar picture of the later 1982 project in Dade County, Florida.[7] Granholm adds that, while many citizens may have been willing to trade privacy for safety [8] and thus did not mind "being watched", some officers were concerned that cameras would be used to monitor the police officer, and that criminals would quickly learn to simply avoid areas within camera range.[9]

[5] *See* Q. Burrows, footnote 4, *supra.*, p. 1103. Burrows describes projects in Hoboken, N.J., Orlean, N.Y., Mt. Vernon, N.Y. and in Times Square. He indicates that all of these first systems were dismantled when found to be ineffective, or when they failed to produce significant numbers of convictions, citing G. Robb, "Police Use of CCTV Surveillance: Constitutional Implications and Proposed Regulations," 13 U. Mich. J.L. Reform 571 (1980). Granholm's article was inspired by the introduction of a significant video surveillance program in Detroit in 1986, and also describes the alleged failure of the early Hoboken and Mt. Vernon projects.

[6] Granholm notes that the Mt. Vernon project produced no convictions, and the Hoboken project led to only one arrest in five years. *See* Granholm, *supra.*, at 688. Burrows reports that the Dade County, Florida project, which was monitored by local volunteers on a 24-hour basis, was discontinued in 1984, with no convictions. Q. Burrows, *supra.*, at 1082.

[7] Burrows notes that although the Dade County project planned to use police employees, community employees, mostly elderly, were used instead, and that the project also experienced significant equipment failure. Burrows, *supra.*, at 1082.

[8] Burrows, *supra.*, at 1103, citing Robb, *supra.*, at 574.

[9] Granholm, *supra.*, at 689. *See e.g.,* Gross v. Taylor, 1997 WL 535872 (E.D.Pa. 1997).

However, Burrows describes subsequent projects in Anchorage, Baltimore, Camden, N.J. (street surveillance of Westfield Acres Housing Projects); Dover (cameras installed in 1993 to monitor the downtown area); South Orange, N.J. (seven cameras monitored by police station personnel); Heightstown, N.J. (cameras installed to monitor trouble spots in housing project); Los Angeles (privately funded program using cameras mounted on apartment buildings to monitor adjacent streets, and using volunteers); Virginia Beach (ten low light sensitive cameras on street light poles at busy beach areas); Tacoma, Boston, Kinston, N.C., Memphis, San Diego's Balboa Park, Ft. Lauderdale, and the Ybor City district of Tampa, Florida.[10] He reports that many of these projects can be described as successful in producing arrests and convictions, reducing criminal activity, and that they can be managed in ways that minimize the risk of intrusive surveillance or taping.[11]

According to several legal writers, the criticism of these projects is not that they cannot be implemented so as to withstand constitutional challenge, but that they are costly and ineffective in bringing about arrests and convictions, and that they add to the negative image of policing by creating a "big brother is watching you" environment on city streets, and places of public accommodation and employment. Privacy concerns are supported by the citation of cases, as well as newspaper accounts of the abusive use of

---

[10] Burrows, *supra.*, at 1106-1108. For descriptions of the Ybor City project, *see* R. Danielson, "Police Cameras May Soon Scan Ybor," St. Petersburg Times, May 24, 1996, 1996 WL 7117611; I. Hathaway, "Decision Delayed on Video Surveillance in Ybor," Tampa Tribune, May 24, 1996, 1996 WL 10228767; and R. Danielson, "Smile, Ybor: You're on Crime Camera," St. Petersburg Times, June 7, 1996, 1996 WL 7119969.

[11] Burrows, *supra.*, at 1122-24. Among cited examples of widely publicized successes are the use of video in the apprehension of the suspects involved in the bombing of the Oklahoma City Federal Building; the Bugler case, in which video surveillance helped police apprehend two boys who murdered a two year old child; the thirty percent drop in crime in Boston housing projects, and significant arrests in Camden, N.J., Memphis, Tennessee, and Tacoma, Washington, as a result of the installation of video surveillance technology. He notes that cities may discourage the unauthorized or abusive use of video by simply avoiding the use of tapes, or recycling them after a certain number of hours.

surveillance technology by police and private security.[12] Finally, commentators cite the recent exploitation of police video for profit as a reason for limiting the use of video surveillance and the video-taping of police activity.[13]

It may be that the interest in video surveillance has persisted because of its growing use in foreign countries. Burrows reports that England has installed more than 150,000 cameras, in more than 75 cities, in response to rising street crime. However, he also reports that many video clips are sold as "bootleg films" on the pornography market. Similar accounts are described in France, where police are given broad powers to install street video surveillance, and in Australia, Ireland and Scotland.[14]

In sum, the history of video surveillance has reaffirmed the common sense notions that all law-abiding citizens are vitally interested in efforts to reduce street crime, crimes in places of public accommodation and other vulnerable places (*e.g.*, ATM machines).[15] However, these same citizens are

---

12 Burrows at 1110, citing, *e.g.*, Doe v. B.P.S. Guard Serv., Inc., 945 F.2d 1422 (8th Cir. 1991)[Security guards filming of fashion models undressing back stage at convention center]; *cf.* Oregon v. Owczarzak, 766 P.2d 399 (Ore. App. 1988), Michigan v. Dezek, 308 N.W.2d 652 (Mich. App. 1988), and Michigan v. Hunt, 259 N.W.2d 147 (Mich. App. 1977)[Police video surveillance of public restrooms]; and newspaper accounts of police abuses of surveillance video. In 1972, Justice Douglas dissented from the Supreme Court's decision not to grant a writ of *certiorari* in Williamson v. United States, 405 U.S. 1026, a case in which the federal appellate court had approved the electronic interception of communications between a police informant and the suspected operator of a whiskey still. Justice Douglas observed that, although electronic eavesdropping had been justified as a necessary means of combating organized crime, it was actually used by government agencies, including the Army, to conduct surveillance of United States Senators and Representatives, the ACLU, the NAACP, the Urban League, and college black studies programs and anti-war groups. *See e.g.*, Chicago Lawyers Committee for Civil Rights Under Law, Inc. v. City of Chicago, No. 76 C 1982, 1985 WL 3450 (N.D.Ill. 1985).

13 Examples include television shows that feature police chases and graphic conduct by suspects, and 911 rescues that feature graphic video of serious injury or death. Such graphic video, it is argued, may cause emotional or physical injury to suspects, victims, and their families. *But cf.* Vega-Rodriguez v. Puerto Rico Telephone Company, 110 F.3d 174 (1st Cir. 1997), holding that fear of employees that employer's silent video surveillance of open work areas might be expanded to "restrooms," creating potential privacy invasion, is not ripe for judicial review until there is a factual basis for such concerns.

14 Burrows cites numerous press accounts of the sale of videotapes of criminal activities, and footage from hidden cameras on streets and in shopping malls and public toilets. *See* Burrows, *supra.*, footnotes 156-176, also describing similar concerns in Australia about cameras in public toilets, and in Scotland about private surveillance of couples making love and people undressing in changing rooms.

15 *See, e.g.*, Dunnigan v. Keane, 137 F.3d 117 (2d Cir. 1998)[State court may admit videotape from bank ATM CCTV to identify assailant who robbed plaintiff and then attempted to use her ATM card, so

worried about the unethical use (viewing, sale, etc.) of surveillance video by police and private security, its inherently indiscriminate and invasive character,[16] and whether, in any event, the cost of broad-scale video surveillance projects will be justified by meaningful increases in arrests and convictions, and a generally significant decrease in criminal activity.[17]

Federal Law

The right of privacy is based in both constitutional law and common law.[18] As a constitutional right, it derives from the First, Third, Fourth, Fifth, Ninth, and Fourteenth Amendments, and from specific provisions of state constitutions.[19] In **Katz v. United States**,[20] the Supreme Court held that the

long as identification is reliable].

16 *See* opinion of Judge Richard Posner in United States V. Torres, 751 F.2d 875 (7th Cir. 1984), permitting the use of "targeted" surveillance video only when the need for surveillance of criminal activity outweighs concerns for privacy; *in accord*, United States v. Biasucci, 780 F.2d 504 (2d Cir. 1986)[Affidavit supported use of video surveillance]. Granholm notes, however, that these cases involved surveillance of private premises, not public streets. *See* Granholm, *supra.*, at footnote 25.

17 *See* L. Linden, "City of Oakland Will Not Use Street Surveillance Cameras," 110 Los Angeles Daily Journal, No. 182, p.3, September 19, 1997, noting a 3-1 vote of the Oakland City Council Public Safety Committee not to proceed with a plan for 50 video cameras to scan streets with zoom lenses. Noting the Council's opinion that such surveillance was legal, the article emphasizes that the ACLU, merchants, and local media had described the plan as Orwellian and a violation of the California Constitution's explicit right of privacy. The article also noted one committee member's opinion that the cost of the cameras could be used to pay for more police officers.

18 Electronic surveillance is not unconstitutional *per se*. *See* U.S. v. Martinez, 498 F.2d 464 (6th Cir. 1974), *cert. denied*, 419 U.S. 1056. See 18 U.S.C.A. §2510. *See also* State v. Diaz, 706 A.2d 264 (N.J.Sup. 1998)[Neither federal or state constitutions are implicated by parents' arrangement with private firm to install a videotape surveillance system in their home to record the conduct of a "nanny" hired to care for their children, citing United States v. Jacobsen, 466 U.S. 109 (1984)].

19 *See, e.g.*, Vega-Rodriguez v. Puerto Rico Telephone Company, 110 F.3d 174, 183 (1st Cir. 1997), holding that while employee surveillance by public employers raises Fourth Amendment concerns, the Ninth Amendment and Fourteenth Amendment cases do not support a cause of action precluding video surveillance of work areas. The court held that the Fourteenth Amendment privacy rights cases generally protect the autonomy of the individual in making significant personal decisions relating to marriage, contraception, family relationships, and the like.

20 389 U.S. 347 (1967).

government's electronic interception of the defendant's conversation in a telephone booth violates his right of privacy, if the defendant had an actual (subjective) expectation of privacy, and that expectation is one that society would recognize as reasonable.[21] This *subjective and objective* test has continued to be the theoretical benchmark in video surveillance cases,[22] but *post-***Katz** cases substantially weaken the expectation of privacy outside the home.[23] Indeed, Burrows and Granholm conclude that the Fourth Amendment is generally not supportive of a constitutional challenge to silent video surveillance of public streets, sidewalks, and parks, because persons do not have a reasonable expectation that they will be free of observation in such public settings.[24]

Granholm argues however that a citizen has some reasonable expectation regarding the extensiveness of technology used to observe her even in public places. Thus, she might have a reasonable expectation that the technology used to observe her in public places would not be so intrusive as to

---

[21] Both Granholm and Burrows note that Katz refused to limit search and seizure protections to cases of physical intrusion, holding instead that the Fourth Amendment protects people, not places. *See* Granholm, *supra.*, footnote 24. Canada has recognized Katz in its interpretation of its own constitutional search and seizure law, holding that, where an individual has a reasonable expectation of privacy, the Charter of Rights and Freedoms would prohibit an unrestricted, warrantless use of surveillance video. *See* Santiago Wong v. Her Majesty the Queen, 3 S.C.R. 36 (1990).

[22] *See* People v. Smith, 360 N.W.2d 841 (1984)[defendant's reasonable expectation of privacy will be determined by totality of circumstances], cited in Granholm, *supra.*, at footnote 26.

[23] Burrows, *supra.*, citing Dow Chemical Co. v. United States, 476 U.S. 227 (1986)[aerial photography by EPA of company's complex]; Texas v. Brown, 460 U.S. 730 (1983)[police officer's use of flashlight to illuminate inside of motorist's car during routine driver's license checkpoint]; Florida v. Riley, 488 U.S. 445 (1989)[aerial surveillance of greenhouse]; California v. Ciraolo, 476 U.S. 207 (1986). In Ciraolo, a 5-4 majority of the Supreme Court held that, although defendant had erected a ten foot fence around his back yard with the intent to obstruct a view of his marijuana growing activity, officers who observed his plants while flying in a private plane at an altitude of 1000 feet did not violate defendant's reasonable expectations of privacy. The court held that the Fourth Amendment protection of the home was never meant to preclude observations that may be made by law enforcement officers from public thoroughfares. Thus, a homeowner's steps to restrict some views does not preclude an officer's observations from a "public vantage point where he has a right to be and which renders [defendant's] activities clearly visible." Defendant's subjective expectation of privacy was therefore not objectively reasonable. 476 U.S., at 213-14.

[24] Granholm, *supra.*, at 694-95; Burrows, *supra.*, at 1090.

focus upon the letter she is reading, or the movement of her lips, or the recording of her words as she walks with a companion.[25] Granholm's argument is based upon her reading of the "plain view" doctrine search and seizure cases.[26] She argues that – although courts have held a view open to outsiders mitigates the suspect's reasonable expectation of privacy – reliance on the plain view doctrine is misplaced where video surveillance includes enhancement features such as telescopic lenses, or film recording devices.[27] Granholm insists that the plain view doctrine is based upon the premise that the discovery of the evidence in question is inadvertent. She then reasons that, where an enhanced video device is deployed to observe activity, the observation is "intrinsically advertent, adverse, and intrusive."[28] However, this aspect of Granholm's argument predates Supreme Court decisions approving aerial searches in drug cultivation cases.

L. R. Willson and Sons v. Occupational Safety & Health Review Commission,[29] considers both the issue of expectation of privacy and Granholm's concern for 'enhancements' such as zoom lenses. In **Willson**, the Secretary of Labor cited the company for OSHA violations after discovering

[25] Granholm, *supra.*, at 695. Granholm argues that this limitation of video and audio surveillance is the essence of a reasonable application of the Supreme Court's decision in Katz.

[26] Citing Coolidge v. New Hampshire, 403 U.S. 443 (1971), limiting the doctrine to situations where police seize an object pursuant to a prior, valid search, *i.e.*, pursuant to a warrant, or a judicially recognized exception to the warrant requirement. Granholm, *supra.*, at 697 and footnotes 42 and 43.

[27] Granholm's distinction has merit. In Vega-Rodriguez v. Puerto Rico Telephone Company, 110 F.3d 174 (1st Cir. 1997), the court observed that arguments justifying video surveillance of streets emphasize the constitutional parity between observations made with the naked eye (by an officer who could be assigned to the street) and observations recorded by an openly displayed video camera having no greater range than the officer's naked eye.

[28] Granholm, *supra.*, at 697. She explains that, if a video camera can zoom in to focus on facial expressions, a license plate, etc., the camera's capability exceeds the senses of the policeman on the beat, and any argument that the camera is simply an extension of the policeman is a flawed argument. She cites People v. Fly, 110 Cal. Rptr. 158 (Cal. App. 1973)[holding that officer's observation of marijuana growing in defendant's enclosed yard through a telescope was s search because the officer had "wedged" himself between two buildings and thus had assumed an unusual vantage point. *In accord, see* United States v. Cuevas-Sanchez, 821 F.2d 248 (5th Cir. 1987).

[29] 134 F.3d 1235 (4th Cir. 1998).

that employees were working on structural steel more than 80 feet above the ground, without the benefit of 'fall protective devices' mandated by 29 C.F.R. § 1926.750(b)(1)(ii). The violation was documented by an OSHA compliance officer who observed Willson workers from the window of a room at a hotel across the street from the worksite, using a "16" power camera lens.[30] Upholding an Administrative Law Judge's admission of the videotape at an evidentiary hearing, the Court of Appeals observed that: "Although surveillance is a type of search that can violate Fourth Amendment protections if performed unreasonably, [the compliance officer's enhanced] observations were not unreasonable." The court held that, since the video disclosed only that which was easily observable by anyone on one of the hotel's upper stories, the employer had no reasonable expectation of privacy. Citing Secretary of Labor v. Concrete Constr. Co.,[31] the Court explained the 'reasonable expectation of privacy rule' in context by observing that there is no constitutional violation when an OSHA inspector makes observations of areas on commercial premises "...that are out of doors and not closed off to the public."[32]

Granholm's second argument is that mass citizen surveillance should be unconstitutional because it lacks the precondition of reasonable suspicion found in drug testing and sobriety checkpoint cases,[33] or the justification for

---

[30] The discovery was initially made by an Assistant Secretary of Labor for Occupational Safety and Health, from his room at the hotel. After observing the workers, the Assistant Secretary telephoned the local OSHA compliance officer who received permission from the hotel to videotape the worksite from the window of the Assistant Secretary's room. The compliance officer then visited the worksite, presented his credentials and interviewed the two employees. 134 F.3d 1235, 1237.

[31] 15 O.S.H.C. 1614 (1992).

[32] 134 F.3d 1235, 1238. The court noted the use of a high powered lens in shooting the videotape, but found that the employer had left the worksite open to observations from vantages outside its control, and thus concluded that a sustained view from a hotel across the street from the construction site was not an unreasonable intrusion into the employer's 'private' space. The court also held that the use of the video camera did not violate Section 8(a) of the Act requiring that an inspector present his credentials before 'inspecting' a site, or the employer's 'walkaround' rights under Section 8(e) of the Act.

mass searches at airports and government buildings. Granholm argues that the cases which allow governmental mass searches at airports and government buildings are based upon the presence of proven present risks of violence in these settings,[34] not present in general surveillance scenarios.[35] She concludes that the undifferentiated threat presented by general crime statistics does not justify the use of highly enhanced surveillance technology. Indeed, she explains, the actual settings in which video surveillance is frequently used are not inner-city high crime areas where the safety of poor people is threatened, but rather areas such as shopping malls, and upscale entertainment districts (*e.g.*, Bricktown, Detroit, and Ybor City, Tampa) where the intent is to protect suburban shoppers, and the economic well-being of store and club owners.[36]

---

[33] Regarding the constitutionality of government checkpoints set up to detect drunk drivers, see Michigan Dept. of State Police v. Sitz, 496 U.S. 444 (1990).

[34] Citing Downing v. Kunzig, 454 F.2d 1230 (6th Cir. 1972); U.S. v. Lopez, 328 F.Supp. 1077 (E.D.N.Y. 1971); U.S. v. Bell, 335 F.Supp. 797 (S.D.N.Y.), aff'd 464 F.2d 667 (2d Cir. 1971), *cert. denied*, 409 U.S. 991 (1972); and federal regulations at 14 C.F.R. 121.538, and 14 C.F.R. 107.123 (1987) requiring air carriers to use screening devices designed to deter passengers from carrying explosives or weapons aboard an aircraft, or to allow unauthorized vehicles access to air operations areas.

[35] Citing Jacobsen v. Seattle, 658 P.2d 653 (Wash. 1983)[The danger posed by patrons at a rock concert is far less than that posed by the threat of terrorist bombings of courtrooms and attempts to hijack airplanes]; Collier v. Miller, 414 F.Supp. 1357 (S.D.Tex. 1976)[Searching university sports arena patrons did not fall under the courthouse or airport exceptions to a warrant requirement]. Granholm acknowledges cases to the contrary, citing Jensen v. City of Pontiac, 317 N.W.2d 619 (Mich. 1982)[Search of patrons entering a stadium is justified by threat of harm by unknown assailants throwing container-type objects]. *And see* Norwood v. Bain, 143 F.3d 843 (4th Cir. 1998)[Approving videotaping of license check of motorcycle riders seeking to enter fairgrounds for motorcycle rally, based upon reliable reports of potential violence involving rival clubs].

The distinction is explained in Wheaton v. Hagan, 435 F.Supp. 1134 (N.D.N.C. 1977). Citing United States v. Edwards, 498 F.2d 496 (2d Cir. 1974), and United States v. Moreno, 475 F.2d 44 (5th Cir. 1973), in addition to the cases cited by Granholm, the court observed that the factors advanced in the airport and courthouse search cases are: the public necessity for warrantless searches, the efficacy of the search, and the nature of the intrusion involved. The public safety factor was met in the seminal cases by the documentation of bombings and other acts of violence in the late 1960's and early 1970's. In contrast, the court explained, as unruly as patrons of arenas and coliseums might be, the dangers posed by these actors are substantially less than the dangers which justified suspending the warrant requirement in courthouse and airport cases. The court was also concerned that in the arena cases, searches tended to be random, singling out certain individuals and thereby exposing them to a stigma or embarrassment. The court held that, unless alternatives to random searches were adopted, a search of an arena patron should comply with the standard announced in Terry v. Ohio, 392 U.S. 1, 21 (1968)[The searching officer must be able to point to specific and articulable facts which...together with rational inferences...(would) reasonably warrant intrusion]. 435 F.Supp. 1134, 1145-46.

[36] Granholm, *supra.*, at 706.

Burrows suggests that attempts to prevent the reasonable use of video surveillance of public places on the ground that such surveillance violates federal privacy concepts are also likely to be unsuccessful. Although he reminds us of the importance of the Supreme Court's decision in **Griswold v. Connecticut**,[37] and its progeny, he suggests that members of the current court have rejected the right of privacy in public places when balanced against the state's interest in deterring criminal activity.[38] The limited precedent dealing with the expectation of privacy is in the context of the video surveillance of suspected criminal activity.

In these cases, federal courts have found some expectation of privacy in business premises, or within buildings, but have upheld video surveillance orders. *See, e.g.,* **United States v. Mesa-Rincon**.[39] These cases deal with the intrusive nature of video surveillance in situations where there is some legitimate expectation of privacy, and where, therefore the need for surveillance must be justified.[40] Such surveillance intrusions, Burrows notes,

---

[37] 381 U.S. 479 (1965).

[38] Citing W. Rehnquist, "Is An Expanded Right of Privacy Consistent with Effective Law Enforcement," 23 Kan. L. Rev. 1 (1974). *And see* Laird v. Tatum, 408 U.S. 1 (1972)[affirming refusal of an injunction preventing army officials from engaging in covert surveillance of civilian political activities where meetings were public]. Burrows, *supra.*, at 1094.

[39] 911 F.2d 1433 (10th Cir. 1990) [authorizing Secret Service installation of television camera to film defendants' counterfeiting operation]; United States v. Wilson, 116 F.3d 1066 (5th Cir. 1997)[Video surveillance of undercover drug purchases from gang members]; *Cf.* U.S. v. Pui Kan Lam, 483 F.2d 1202 (2d Cir. 1973), *cert. denied*, 415 U.S. 984 [Defendants had no justifiable expectation of privacy while in the house of complete strangers to which they had gained entry by false representations].

[40] *Cf.* Vega-Rodriguez v. Puerto-Rico Telephone Company, 110 F.3d 174 (1st Cir. 1997), defining the scope of a public employer's right to conduct disclosed silent video surveillance of open employee work areas. In Vega-Rodriguez, a quasi-public employer installed, over employee objection, a silent video surveillance system to record all employee activity in open work areas. The videotapes were stored and could be viewed with the permission of a designated company official. Citing Oliver v. United States, 466 U.S. 170 (1984), and O'Connor v. Ortega, 480 U.S. 709 (1987), the court recognized that public employees may be protected against unreasonable search and seizure if the challenged conduct infringes a reasonable expectation of privacy. That protection must, however, be both subjectively demonstrable and objectively reasonable under the circumstances. Generally, the employee's expectation of privacy is objectively reasonable as to his exclusive private office, desk and file cabinets containing personal matters not shared with other workers. In contrast, there is no reasonable expectation of privacy against video surveillance of open work areas, unenclosed locker areas, or desks, files, and the like subject to shared access among employees, especially where the employer discloses its

are also the concern of Title III of the Omnibus Crime Control and Safe Streets Act of 1968, as amended.[41] Unfortunately, the federal courts appear to be divided on the application of the Act's requirements to targeted silent video surveillance, where justifiable expectations of privacy might exist.[42] Thus, federal law in this area remains less than fully conclusive.

---

use of surveillance. Citing State v. Taketa, 923 F.2d 665 (9th Cir. 1991); Schowengerdt v. United States, 944 F.2d 483 (9th Cir. 1991); Sheppard v. Beerman, 18 F.3d 147 (2d Cir. 1994); American Postal Workers Union v. United States Postal Service, 871 F.2d 556 (6th Cir. 1989); Thompson v. Johnson County Community College, 930 F.Supp. 501 (D. Kan. 1996); and Gross v. Taylor, 1997 WL 535872 (No. Civ. A. 96-6514, 1997)[holding that public police officers did not have an objectively reasonable expectation of privacy or non-interception while on duty in a patrol car, and thus interception of their conversation by employer's open and visible rear seat microphone would not violate 18 U.S.C. §2510, or Fourth Amendment protections].

41 18 U.S.C., §§ 2510-2521, as amended by the Electronic Communication Privacy Act of 1986, Pub..L. 99-508, P.L. 103-414, and P.L. 105-153 (1997). *See* United States v. Bailey, 607 F.2d 237 (9th Cir. 1979), holding that Title III is constitutional under the Fourth Amendment, citing Hackett v. United States, 429 U.S. 837 (1975). The Act requires application for a court order to intercept communications in connection with the investigation of enumerated crimes, but does not specifically subsume the use of silent video surveillance. See Ricks v. State, 537 A.2d 612, 613-14 (Md. 1988). The purpose of the Act is to protect individual privacy while permitting limited governmental surveillance in accordance with uniform standards, when necessary for effective law enforcement. *See, e.g.*, Application of U.S. Authorizing Interception of Wire Communications, 413 F.Supp. 1321 (E.D.Pa. 1976); Dowd v. Calabrese, 101 F.R.D. 427 (D.D.C. 1984). For an excellent discussion of Title III, *see* K. Springer, "In God We Trust: All Others Who Enter This Store Are Subject to Surveillance." 48 Fed. Comm. L.J. 187 (1995).

It is noteworthy that, while §8 of Canada's Charter of Rights and Freedoms (protecting against unreasonable search and seizure), is similarly silent as to video surveillance, the Supreme Court of Canada has interpreted §8 to cover both audio and video surveillance. *See* Santiago Wong v. Her Majesty the Queen, 3 S.C.R. 36 (1990), citing R. v. Duarte, 1 S.C.R. 30 (1990)[The court observed: "In Duarte, this court held that unauthorized electronic audio surveillance violates §8 of the Charter. It would be wrong to limit the implications of that decision to that particular technology. Rather, *what the court said in Duarte must be held to embrace all existing means by which the agencies of the state can electronically intrude on the privacy of the individual, and any means which technology places at the disposal of law enforcement authorities in the future*," (emphasis added), citing Brandeis, J. in Olmstead v. United States, 277 U.S. 438 (1928)]; *and see* United States v. Cuevas-Sanchez, 821 F.2d 248, 250-52 (5th Cir. 1987).

42 *Compare* United States v. Mesa-Rincon, *supra.*; United States v. Cuevas-Sanchez, *supra.*, 821 F.2d 248, 250-52 (5th Cir. 1987)[defendant had reasonable expectation of privacy, under California v. Ciraolo, 476 U.S. 207 (1985), against government's video surveillance of his fenced-in back yard through use of camera installed to indiscriminately record all of his activity; held however, that government complied with Title III of OCCSSA in obtaining surveillance order]; United States v. Biasucci, 768 F.2d 504 (2d Cir. 1986); and United States v. Torres, 751 F.2d 875 (7th Cir. 1984); with United States v. Taketa, 923 F.2d 665 (9th Cir. 1991)[silent video taping does not come within provisions of Title III]; In Re Order Authorizing Interception of Oral Communication & Video Surveillance, 513 F.Supp. 421 (D. Mass. 1980)[Title III not applicable to silent video surveillance]; U.S. v. Foster, 985 F.2d 466 (9th Cir. 1993), *on rehearing*, 17 F.3d 1256 [videotaping of defendant did not violate Electronic Privacy Act where a number of persons were present and with the consent of the owner of the premises]. State v. Diaz, 706 A.2d 264 (1998), emphasizes the Seventh Circuit's comment in United States v. Torres, *supra.*, that "[Of course], it is anomalous to have detailed statutory regulation of bugging and wiretapping, but not of television surveillance, in Title III, and detailed statutory regulation of television surveillance of foreign agents, but not of domestic criminal suspects, in the Foreign Intelligence Act....But judges are not authorized to amend statutes, or bring them up to

State Law

Concepts of privacy have been fashioned by the states in constitutional provisions and judicial pronouncements. Several states, including Oregon, Pennsylvania, Hawaii, Montana, Illinois, California, Alaska, Florida, New Hampshire, and Michigan have explicit constitutional protections of privacy, some of which limit search and seizure, including wire and electronic communications surveillance which might be permitted by U.S. Supreme Court precedent.[43] However, several states have permitted video surveillance when supported by legitimate public interest in newsworthy information.[44] Indeed, Burrows notes, the public interest in crime can overcome personal concerns for privacy even in situations where publication of videotaped accounts cause emotional upset.[45]

date"; *Cf.* United States v. Andonian, 735 F.Supp. 1467 (C.D.Cal. 1990)[legislative history of 1986 amendments to Title III suggests that statute would apply to audio but not video portion of a surveillance]; aff'd and remanded, 29 F.3d 634 (9th Cir. 1994); *cert. denied,* Andonian v. United States, 513 U.S. 1128 (1995). For a discussion of the use of a 'highlighted portion' of a videotape in criminal proceedings, *see* United States v. Hardwell, 80 F.3d 1471, 1492-93 (10th Cir. 1996).

Canada appears to follow Biasucci, and holds that video surveillance of a hotel room would normally be held to be a search, and thus would require a warrant. Santiago Wong v. Her Majesty the Queen, 3 S.C.R. 36 (1990), also citing Stoner v. California, 376 U.S. 483 (1964) regarding the reasonable expectation of privacy of one occupying a hotel room (A majority of the Court in Wong held, on the facts presented, that the search in question, although including video surveillance not authorized by a court, was reasonable based upon legal advice received by police, and where it was arguable that defendant had no reasonable expectation of privacy during "floating" gaming operation. The court held however, that unauthorized, surreptitious electronic surveillance violates §8 of the Charter of Rights and Freedoms where the target of the surveillance has a reasonable expectation of privacy).

43 Burrows, *supra.*, at 1113-1114. Florida's concern for inappropriate private use of silent video has led to the proposal of HB 3709, Chapter 98-415, creating Section 810.14, F.S., prohibiting a person from secretly observing, photographing, or videotaping another person with lewd...or indecent intent, when the victim is in a dwelling, structure, or conveyance that provides a reasonable expectation of privacy. Interest in the passage of the statute was apparently fueled by reports of a case involving female workers at the Apalachicola Times who disclosed that a manager was observing them via a palm-sized video camera installed behind an air conditioning vent in an employee rest room, with live video feed to a monitor in the manager's office. *See* M. Lasswell, "We're Being Watched", Allure Magazine (August, 1998), p.135. Lasswell notes the expanding video-voyeur subgenre on the Internet under the title "Upskirt," that has dozens of sites devoted to the display of pictures of unsuspecting women taken in malls, parks, stores, etc.. taken by concealed cameras.

44 *Id.*, citing Gill v. Hearst Publishing Co., 239 P.2d 636 (Cal. 1952); *reh.* 253 P.2d 441 (Cal. 1953), and DeGregorio v. CBS, 473 N.Y.S.2d 922 (N.Y.Sup. 1984).

45 *Id.*, at 1116-1119, citing Waters v. Fleetwood, 91 S.E.2d 344 (Ga. 1956)[newspaper publication of photographs of murdered fourteen year old girl held newsworthy]; Cape Publications, Inc. v. Bridges,

Targeted video surveillance may be permitted under state law adopting the **Katz** standard. In **Ricks v. Maryland**,[46] the Baltimore police department employed surreptitious, nonconsensual video surveillance, pursuant to court order, as part of an extensive narcotics investigation of premises allegedly being used by defendants as a "processing house" or "cut house" where controlled dangerous substances were diluted and packaged for street sale.[47] Following the arrest of defendants based upon a search warrant, the appellate court upheld the court-ordered surveillance.[48] The court noted defendants' admission that video surveillance was not regulated by the federal Omnibus Crime Control and Safe Streets Act of 1968, after which the Maryland wiretap statute was modeled,[49] and that the Maryland statute did not expressly contemplate video surveillance.

The court held, therefore, that silent video surveillance of suspected criminal activity was not proscribed by the Maryland Wiretap and Electronic Surveillance Act.[50] As to defendants' Fourth Amendment argument, the court

---

423 So.2d 426 (Fla. 1982)[newspaper publication of photograph taken of rape victim at scene of crime shortly after she was raped by former husband].

46 520 A.2d 1136 (Md. App. 1987); *aff'd*, 537 A.2d 612 (Md. 1988).

47 Police were allowed to install a small video camera into the ceiling of the apartment to record the illegal activities, after a showing that alternate investigative methods had been tried and failed, or were too dangerous to undertake. 537 A.2d at 613, 615. (The court reviewed in detail the showing required under Section 2516-18 of the Federal Act).

48 Defendants argued that the surveillance violated both the Maryland Wiretap and Electronic Surveillance Law, and the Fourth Amendment to the U.S. Constitution. 520 A.2d 1136, 1138.

49 537 A.2d at 613-14.

50 Citing United States v. Biasucci, 786 F.2d 504 (2d Cir. 1986), *cert. denied*, 479 U.S. 827 (1986)[video surveillance of "business" offices of loan-sharking operation]; United States v. Torres, 751 F.2d 875 (7th Cir. 1985)[video surveillance of terrorist "safe houses" used to assemble bombs permitted on showing that audio devices alone might be neutralized by defendants playing music, using code, or assembling bombs in silence]; In the Matter of an Application for an Order Authorizing Interception of Oral Communications and Video Surveillance, 513 F.Supp. 421 (D. Mass. 1980); and People v. Teicher, 422 N.E.2d 506 (N.Y. 1981)[permitting court order allowing installation of camera in office of dentist suspected of sexually assaulting female patients (*see* R. Hochberger, "Appellate Division Approves Video Surveillance by Police," 37 New York Law Journal, February 25, 1980)].

reasoned that the proponent of a motion to suppress has the burden of proving that the video surveillance in question violates a legitimate expectation of privacy in the invaded place.[51] Citing **Smith v. Maryland**,[52] the court held that defendant must demonstrate, by his conduct, that he has exhibited a subjective expectation of privacy (that he seeks to preserve something as private), and that his expectation is one that society is prepared to recognize as reasonable (that is, whether the defendant's expectation, viewed objectively, is justifiable under the circumstances).[53]

Some states have constitutional provisions which arguably prohibit general police use of powerful video street surveillance cameras with zoom lens capability, or other intrusive surveillance.[54] Burrows cites **Hawaii v. Bonnell**,[55] holding that the video surveillance of an employee break room by police without a warrant (to investigate alleged gambling operations) violated

---

[51] 537 A.2d at 619, citing Katz v. United States, *supra.*, and Rakas v. Illinois, 439 U.S. 128 (1978).

[52] 442 U.S. 735 (1979).

[53] 537 A.2d at 619. Finding that defendants may have had a reasonable expectation of privacy under the facts of the case, the court of appeals held that the video surveillance was conducted in accordance with Fourth Amendment requirements and consistent with the required showings under Title III of the OCCSSA. 537 A.2d at 613, 620-21. The Ricks standard is explained in McCray v. State of Maryland, 581 A.2d 45 (Md. App. 1990). In McCray, the defendant was suspected of procuring false driver's licenses for persons whose licenses had been suspended or revoked. As a part of their investigation, police conducted a warrantless video surveillance of defendant, videotaping him crossing the street to a state motor vehicle administration office. The court held that such surveillance did not implicate the privacy concerns evident in Ricks, because the video surveillance of defendant took place only when he was crossing the street and entering the MVA office in full public view. Citing Katz, and Note, "Police Use of CCTV Surveillance: Constitutional Implications and Proposed Regulations," 13 U. Mich. J.L.Ref. 571 (1980), the court held that a person does not have a reasonable expectation of privacy when he is walking along public sidewalks, streets, or parking lots, or in a similar location in full public view. 581 A.2d, at 47-48. *See also* State v. Diaz, 706 A.2d 264 (N.J.Sup. 1998)[New Jersey's Wiretap Act, which is modeled after Title III of federal Omnibus Crime Control Act and Safe Streets Act, does not subsume silent television surveillance, and the legislative history of the federal legislation indicates that the exclusion was deliberate. The admissibility of a videotape with sound recording in a criminal proceeding is, however governed by the warrant provisions of the New Jersey statute].

[54] *See* L. Linden, *supra.*, footnote 16, citing the concern of Oakland City Officials that such surveillance would violate the California Constitution.

[55] 856 P.2d 1265 (Hawaii, 1993).

the Hawaii Constitution.[56] Burrows also emphasizes Montana's requirement of a compelling governmental interest to justify excessively intrusive surveillance.[57] However, where video street surveillance is limited in its intrusiveness, some legislatures have proposed that its use in reducing traffic violations or crime is justified.[58]

Canada has defined this intrusiveness facet of the constitutionality of video surveillance in precise terms. The Supreme Court has rejected a "risk analysis" which would permit surveillance if an assessment of the person's reasonable expectation of privacy were made to rest on a consideration whether he "courted the risk of electronic surveillance." Rather, in **R. v. Duarte**,[59] the court defined the reasonable expectation of privacy in given circumstances by asking whether – by the standards of privacy that persons can expect to enjoy in a free society – the state should not be allowed to engage in the surveillance questioned without prior judicial authorization. This interpretation suggests that constitutional protections against unreasonable search and seizure must embrace an awareness of advances in the science and technology available to government. The court speculates that, given the

---

56 The Court noted that fifty video tapes with 1200 hours of footage disclosed only one minute of conduct which might reflect gambling activity. More important, the court held that the Hawaii constitution protects legitimate expectations of privacy wherever the individual may go. Thus, even in a public park, an individual may have an expectation of privacy that should not be invaded by warrantless video surveillance absent exigent circumstances. Indeed, the court held that the justification required for video surveillance should be higher than that required for audio surveillance. Id., at 1273.

57 State v. Brown, 755 P.2d 1364 (Mont. 1988), approving consensual warrantless monitoring of face-to-face conversation, but observing that privacy might preclude such interception where none of the participants has consented to the surveillance.

58 *See* Maryland House Bill 391, SIA Second Annual Report on CCTV for Public Safety, Appendix 8, p. 153 [proposed legislation for use of video cameras to capture image of automobile and license plate entering intersection after traffic signal has turned red]; Illinois House Resolution 62, SIA Report, Appendix 12, p. 211[enhancing video surveillance technology used by merchants by installing measurement device within picture to better show criminal's height and size].

59 1 S.C.R. 30 (1990).

advanced state of surveillance technology, a "risk analysis" would set a meaningless standard for privacy.[60]

Employer Use of Video Surveillance in Work Areas

The Seventh Circuit's opinion in **Brazinski v. AMOCO Petroleum Additives Company**,[61] presents a thought-provoking discussion of employee expectations of privacy and employer concern for employee misconduct. In **Brazinski**, eight female workers (at an AMOCO chemical laboratory) challenged in state court their employer's installation of a television camera in the ceiling of a locker room used by female employees to change from their street clothes into work clothes. The company explained that its CCTV installation was for the purpose of documenting improper (presumably sexual) activity by a certain male employee and a certain female employee in this locker room during working hours. However, seeking to avoid a determination of the merits, the company argued that the suit implicated the company's collective bargaining agreement with plaintiffs' union, and thus was a suit under Section 301 of the Taft-Hartley Act, 29 U.S.C. § 185. The suit was successfully removed to federal district court, and summary judgement was entered in favor of the company because plaintiffs failed to file a grievance under the bargaining agreement. Affirming summary judgement against the union plaintiffs, the Seventh Circuit nonetheless

[60] Santiago Wong v. Her Majesty the Queen, 3 S.C.R. 36 (1990), citing also Amsterdam, "Perspectives on the Fourth Amendment", 58 Minn. L. Rev. 349, 402 (1974). It may be said that the U.S. Supreme Court agrees with the Canadian court's general concerns, but would apply those concerns only to restrict surveillance which violates a truly reasonable expectation of privacy. In California v. Ciraolo, *supra.*, [approving warrantless surveillance by police of defendant's back yard by private plane, on tip about defendant's marijuana growing activities], the majority recognized Justice Harlan's concern in Katz v. United States, that future electronic developments would extend the potential for electronic interference with private communications. However, the majority held, such concerns are not aimed at simple observations from a public place. 476 U.S., at 214.

[61] 6 F.3d 1176 (7th Cir. 1993).

rendered an opinion on the 'supplemental' claim of one non-union plaintiff[62] that the company's use of CCTV in the female employees' locker room subjected her to being taped in a state of undress.

The court observed that state tort law might support a claim that a 'well-motivated but unavoidably indiscriminate effort at [video] surveillance is actionable by a person [who is] not the target of the surveillance, [but] who accidentally wanders into the scene and is photographed or recorded [in a state of undress]."[63] However, the court cautioned, if the method of surveillance chosen is the least indiscriminate possible for achieving a lawful and important objective, the 'stranger' whose privacy is incidentally and accidentally compromised might not have a cause of action.[64] If a cause of action were to be recognized, the court observed, plaintiff would be required to demonstrate that she was seen live by a human being – either a person monitoring the camera, or viewing a tape – or, at the very least that she was in the place under surveillance so that, if the equipment was manned, she would have been seen or heard.[65] *See also* **Vega-Rodriguez v. Puerto Rico Telephone Company**, *supra., footnote 40,* 110 F.3d 174 (1st Cir. 1997), observing that announced, silent video surveillance of public employee work areas open to all employees is less intrusive than a physical search that intrudes into employee desks, drawers, filing cabinets, or other enclosed spaces, and does not intercept private conversations between employees.

---

[62] Jones, the non-union plaintiff, was in fact an electrician employed by another company – but engaged in work on defendant's premises.

[63] 6 F.3d 1176, 1183.

[64] *Citing* Birnbaum v. United States, 588 F.2d 319 (2d Cir. 1978). The court noted that this situation could arise, for example, where an innocent person visits an apartment that is under police surveillance.

[65] Finding that plaintiff introduced no evidence either that she was in the locker room during periods of CCTV surveillance, or that the camera was aimed inside the locker room (the employer argued that the camera was aimed at all times at the entry door to the locker room), the court affirmed dismissal of plaintiff's complaint.

Union activity receives special protection from video surveillance under § 8(a)(1) of the National Labor Relations Act, 29 U.S.C. § 158(a)(1). Recent cases such as **California Acrylic Industries, Inc. v. National Labor Relations Board**,[66] hold that, in the absence of proper justification (*e.g.*, violence or trespass), the videotaping of union pickets, or union/employee activity during organizing efforts or contested elections has a tendency to intimidate and interfere with the employees' right to engage in concerted activity. Specifically, the court held in **California Acrylic**, the employer may not videotape such activities on the basis of an undifferentiated fear that 'something might happen.'[67] Moreover, even where the employer's videotaping is justified as a lawful precaution against violence, it must be careful not to exceed the necessary boundaries of surveillance activities.[68]

Summary of Constitutional and Privacy Issues

Burrows suggests that serious consideration must be given to the argument that the extension of expectations of privacy to public places so as to preclude video surveillance will, in fact, impede law enforcement efforts to protect the public from crime. He observes that many citizens support public surveillance programs, so long as they comply with the need to prevent abusive use of the technology, or

---

[66] 150 F.3d 1095 (9th Cir. 1998).

[67] 150 F.3d 1095, 1100; *and see* National Steel and Shipbuilding Company v. National Labor Relations Board, 156 F.3d 1268 (D.C. Cir. 1998); Clock Electric, Inc. v. National Labor Relations Board, 162 F.3d 907 (6th Cir. 1998); *Cf.* Overnight Transportation Company v. National Labor Relations Board, 140 F.3d 259 (D.C. Cir. 1998), holding that videotaping by employees who were union supporters but not union members did not violate the Act, where employer could not show that their activities should be attributed to the union under common law agency principles. Where election misconduct is attributable to one of the parties, the Board will overturn the election if the misconduct created such an environment of tension and coercion as to have a probable effect upon employee action at polls and to have materially affected the results of the election. Where misconduct is attributable to third parties, however, the Board will overturn the election only if the misconduct is so aggravated as to create a general atmosphere of fear and reprisal rendering a 'free election' impossible. 140 F.3d 259, 302-03.

[68] *See* Horsehead Resource Development Co., Inc. v. National Labor Relations Board, 154 F.3d 328 (6th Cir. 1998)[In taping beyond front gate of plant, surveillance of union members who were in no way engaged with company employees or property, but were merely talking among themselves or moving to and from picket shack and portable toilet, was unjustified].

videotapes. He advises however, that the warrantless use of video surveillance by police should be limited to public streets where the Supreme Court has held that citizens have no reasonable expectation of privacy.[69] Burrows emphasizes that the right of privacy is a fundamental right in our society, and that the more than 600,000 state and federal law enforcement personnel and 1.5 million private security personnel, with resources in excess of thirty billion dollars, present a force that has already eroded notions of privacy once taken for granted. And, both legal writers and journalists express concern that video surveillance may be used by police to "target" minorities who are stereotyped as more likely to commit crimes, as well as members of unpopular political action groups in the community. Burrows cautions that, in our efforts to reduce crime, we must not trade individual liberties for rigid notions of order. Technology should be used to support arrests only where it is reliable, and aspects of its unreliability or potential abuse must be understood.[70]

Tort Claims Related to the Use of Silent Video: Private Civil Liability and Governmental Immunity

Introduction: Modern tort law has been increasingly concerned with the security of premises, but modern rules both expand and limit duty. In most, if not all jurisdictions, modern duty rules were originally announced by courts or legislatures to limit the 'old' common law's imposition of strict

---

69 Burrows, *supra.*, at 1124.

70 Burrows, *supra.*, at 1125-26. He notes, for example, that digital imaging allows a criminal to be removed from a scene or placed at a scene, and that an expert could not distinguish a copy from the original master tape. He also expresses concern that citizens could access surveillance footage through the Internet on their personal computers. Some jurisdictions are already establishing a structure for bringing together groups having competing concerns about the parameters of the use of video surveillance in targeted retail/commercial areas of the City. *See, e.g.*, Agenda Report, City of Oakland, SIA Second Annual Report on CCTV for Public Safety, Appendix 9, p. 179-80 [but noting decision not to seek opinion of City Attorney unless there is a governmental role in the surveillance program].

Burrows proposes a model statute defining the permissible scope of video surveillance. His statute would provide, *inter alia*, that: All surveillance operators must be trained, professional, certified police or federal agents; Operators should make specific disclosure to targets of surveillance, along with a general public disclosure of the video surveillance activities of police departments to citizens who must then be allowed to submit comments and objections at public hearings; Operators must prove, by a showing of probable cause and compelling governmental interest that video surveillance is necessary and that the least restrictive [sic] method of surveillance will be employed; Targeted surveillance should be permitted only on showings and according to procedures presently required under statutes like Title III, as amended; That under no circumstances shall the contents of any captured video images be exploited for purposes of profit, publication or distribution; and that violations of the statute would lead to the suppression of evidence, criminal penalties, and/or civil remedies. 31 Val. U.L. Rev. 1079, 1133-1138.

Also noteworthy in summary is the suggestion that we continue to give some consideration to the opinion of Justices Powell, Brennan, Marshall and Blackmun in California v. Caraolo, 476 U.S., at 216-226.

liability on innkeepers for loss or damage to a guest's personal property. **McIntosh v. Schops**, 180 P. 593 (Ore. 1919), cited in **Kutbi v. Thunderlion Enterprises, Inc.**, 698 P.2d 1044 (Ore. App. 1985). However, in modifying the 'old' common law rule, modern courts imposed a duty on landowners to exercise reasonable care for the safety of business or public invitees.

The modern rule, summarized in the Restatement 2d, Torts, provides that a landowner that holds land open to the public is subject to liability for physical harm to invitees caused by the accidental, negligent, or intentionally harmful acts of third persons, if the landowner fails to use reasonable care to (a) discover that such acts are occurring or are likely to occur, or (b) adequately warn visitors to avoid such harm or otherwise protect them from it.[71] Because the rule is derived from negligence (fault) principles, and not strict liability theory, liability is 'pegged' to foreseeability of harm. And, because the landowner is not generally required to anticipate that third parties will commit criminal acts, the landowner is subject to liability only where criminal intrusion is reasonably foreseeable.

The rule is usually stated to provide that the landowner –*.e.g.*, landlord – may be negligent, even though the harm to a visitor/invitee – *e.g.*, tenant – is caused by the criminal act of a third person, if the situation is one in which a reasonable landowner would have foreseen the likelihood of criminal intrusion.[72] The landowner/proprietor is not the insurer of the invitee's – *e.g.*, tenant's – safety[73], but is required to exercise reasonable care to protect the invitee from unreasonable risks of which the landowner has superior knowledge. What constitutes reasonable care in a given situation varies with the circumstances, but generally evidence of substantially similar prior

---

[71] Restatement 2d, Torts, Sec. 344 (1965).

[72] *See* Restatement, 2d, Torts, Sec. 302, and Comment e.

[73] *See* Scott v. Harper Recreation, Inc., 506 N.W.2d 857 (Mich. 1993).

criminal acts may be used to demonstrate that the landowner had actual or constructive knowledge of risk of harm to the invitee. The term 'substantially similar' does not mean identical – as, for example, whether a weapon was used – but whether the prior crimes would put a reasonable landowner on notice that visitors, residents, etc. were subject to increased risk of harm. The question is whether the prior activity would have attracted the attention of a reasonably prudent landowner, and caused him to be concerned about the safety of visitors, tenants, etc. **Shoney's Inc. v. Hudson**, 460 S.E.2d 809 (Ga. App. 1995); **Cohen v. Southland Corporation**, 203 Cal. Rptr. 572 (Cal. App. 1984)[Citing cases from Pennsylvania, New York, North Carolina, Oregon, Texas, Massachusetts, and New Jersey]. What is required to be foreseeable is the general character of the event or harm, not the precise nature of the activity or the precise manner of its occurrence.[74]

Balanced against this consideration of the likelihood and severity of harm to visitors, tenants, etc. is the burden to the landowner if he is required to eliminate or reduce the risk. Where reasonable efforts to reduce risk would not place an onerous burden on the landowner, it is more likely that he will be asked to take affirmative steps to reduce the risk of criminal activity that threatens visitors, tenants, etc.[75]

To summarize, until recently courts have been reluctant to impose liability on the landowner/operator of premises for injuries to the landowner's invitees/tenants/customers/students, etc., caused by the criminal act(s) of third parties. However, recently many courts have extended negligence rules to hold supermarkets, restaurants, libraries, schools, summer camps, and other

---

[74] Cohen v. Southland Corp., 203 Cal. Rptr. 572, 576. *Contra.*, *see* Boren v. Worthen National Bank of Arkansas, 921 S.W.2d 934 (Ark. 1996)[Holding that bank is not required to provide security at ATM's and that the fact that apartments, or businesses are in high crime areas does not in itself establish a duty to provide security. The dissent argues that the court should adopt the foreseeable risk rule, and observes that it should be a question of fact whether installation of cameras, or other measures, would have deterred criminal acts that caused plaintiffs' injuries].

[75] Id, at 578.

entities liable for crime-related injuries. Even though the criminal act is, in fact, an intervening act, the landowner's antecedent negligence subjects him to liability if the criminal act was itself reasonably foreseeable. *See* **Nebel v. Avichal Enterprises, Inc**., 704 F.Supp. 570 (D.N.J. 1989).[76]

Application of the Rule in the Context of Video Surveillance: (1) Failure to utilize a video security system: A natural aspect of a modern claim of negligent security is that a landowner's security system does not include available, cost-effective, and popularly utilized video surveillance equipment. In tort law cases, a plaintiff may actually introduce evidence of the 'industry standard' to show negligence. Although departure from the 'industry standard' does not establish negligence *per se*, the benchmark is relevant and admissible. In **Nebel v. Avichal Enterprises, Inc.**, 704 F.Supp. 570 (D.N.J. 1989), a motel patron alleged that Defendant's employees were negligent in failing to provide 'functional and operational closed circuit surveillance cameras and monitors,' in a motel located in a well-known high crime area of Atlantic City, New Jersey. The court held that the obligation of plaintiffs in cases alleging inadequate security is to prove that defendant's negligence was a substantial factor in causing the harm. Keeping in mind the basic rule of law (that, while the criminal act is, in fact, an intervening act, the defendant remains liable if such a criminal act was foreseeable, and the defendant did not exercise reasonable care to reduce the risk of its occurrence), the plaintiff needs to prove, essentially, that a video surveillance system – or other security measures – would likely have deterred the criminal activity that caused plaintiff's injury.[77] *See also* **Morris v. Krauszer's Food Stores, Inc**., 693 A.2d 510 (N.J.App. 1997)(Jury award of damages affirmed where plaintiff's estate introduced expert testimony that, considering foreseeability of robbery, defendant should have increased security measures, including the installation of video cameras[78]).

(2) Use of video to replace security guards or officers: Although the cases are few in number, some observations may be made about the 'reasonableness' of using silent video surveillance to replace or enhance security

---

[76] There is one important limit to this liability. Because the allegation of negligent security in such situations is based upon the invitee's status and relationship with the landowner, the landowner's liability does not extend beyond his premises, and even on premises extends only to those areas within the landowner's control. Thus, for example, a landlord's duty to provide reasonable security to his tenants extends to those areas of the landlord's premises over which the landlord retains control during the lease (common entrances, stairwells, laundry rooms, recreation facilities, etc.).

[77] 704 F.Supp. 570, 580. The court's opinion contains a detailed discussion of the law of proximate causation that need not be detailed here, but that is instructive to the attorney or administrator who desires an in depth discussion of the principle of 'significant factor' analysis that underlies proximate cause theory in cases involving two or more alleged causes of harm.

[78] Specifically, plaintiff's expert testified that a clearly visible closed circuit television camera focused on the area of the cash register, and a barrier to protect employees should have been installed. 693 A.2d 510, 513.

personnel/guards/officers. In **<u>Shoney's, Inc. v. Hudson</u>**, *supra.*, a patron was robbed and injured by an assailant in the parking lot of defendant's restaurant. She alleged that defendant knew of at least four acts of violence at this location within the prior two years, including one shooting of a cashier. Plaintiff then alleged that defendant had initially responded by hiring security personnel during all evening hours, but later discontinued the use of guards and installed silent video cameras near the cash register of its new restaurants. Later, defendant apparently hired guards to observe the premises and escort employees with payrolls on Friday and Saturday nights. Noting that the restaurant was located in the highest crime area of any of defendant's outlets in Savannah, and that defendants had acknowledged the potential for criminal attacks, the court held that an issue of material fact was raised whether defendant provided reasonable security for patrons.[79]

> The opinion may be read to state that a business or public invitee may allege that a landowner (*e.g.*, storeowner)/landlord/school, *etc.* is negligent in discontinuing the use of security personnel – and replacing them with silent video – where the video is ineffectively deployed or located, or where reasonable care calls for the use of security guards or officers. Certainly the court's opinion cautions – if indirectly – against the undifferentiated use of video to replace security personnel, merely to save money, *etc.*

(3) <u>Policies and procedures, and employee training</u>: Where devices are installed as the only security measure, or to enhance security, the landowner must also be careful to follow its own policies regarding installation and use of the technology. **<u>Cohen v. Southland Corporation</u>**, 203 Cal. Rptr. 572 (Cal. App. 1984) is illustrative. The defendant corporation had commissioned a study of its store security, and had embarked on a program of employee training, balancing of lighting (inside and outside stores), *etc.* When a patron was shot by a robber – while the store clerk hid in the back room of the store – the patron alleged that the installation of a security camera at the cash register did not represent adequate security, unless store employees were

[79] 460 S.E.2d 809, 812.

adequately trained, and store interiors and parking lots were properly illuminated, *etc.*[80]

(4) False sense of security: Victims of robbery or assault might also allege negligence where there is evidence that a video security system is improperly designed or maintained, or not monitored. The latter allegation may actually include a claim that a video security system that is represented as monitored, but is in fact not monitored, may create a false sense of security, thereby encouraging visitors/tenants/customers/students to take risks they would not take if they knew the video security system was not monitored. *See, e.g.*, **Kutbi v. Thunderlion Enterprises, Inc**., 698 P.2d 1044 (Ore. 1985). In **Kutbi**, the patron of a motor inn alleged that defendant's employees were negligent – with respect to guest security – when they: (1) duplicated excessive keys; (2) maintained a video security system that was not in working order, and that did not offer a view of patrons' rooms; and (3) 'lulled' patrons into a 'false sense of security' by not disclosing that the video security system was not regularly monitored. While the court's opinion does not comment in detail on plaintiff's third allegation, it implies that defendant prevailed on this issue only because it introduced evidence that the security system was in good working order, and was properly monitored on the night when plaintiff's room was burglarized.[81]

(5) Governmental immunity: The 'public duty doctrine' limits the liability of governmental landowners by generally precluding – under constitutional 'separation of powers' analysis – a judicial imposition of executive branch policy. In other words, a private plaintiff – who is a crime

---

[80] Thus, where plaintiff introduced evidence that the store manager had received no security training – despite the corporation's assertion that it had an extensive program of employee security training – and evidence of inadequate lighting on the night of the incident, summary judgement for the corporation was properly denied. The inference is, of course, that use of a video camera is not *per se* reasonable care where the situation demands additional forms of/or approaches to security.

[81] 698 P.2d 1044, 1048. The court denied Defendant's motion for summary judgement, finding that Plaintiff had introduced evidence sufficient to create a genuine issue of material fact on the questions whether Defendant's employees had made excessive keys and had not changed locks.

victim – could probably not obtain a private monetary award based upon the allegation that a public landowner (*e.g.*, a university) had generally devoted too few resources (too little budget) to (campus) security. Governmental agencies/entities should however be cautious not to play 'fast and loose' with this qualified immunity. Where a visitor/tenant/student enjoys a legal relationship with the landowner, the landowner's duty as landlord, premises operator, *etc.*, makes it vulnerable to allegations of negligent security to the same extent that a private landlord/premises operator/*etc.* would be subject to liability. Indeed, in reality, except as to 'undifferentiated' allegations of negligent security, the public institution, under most Tort Claims Acts, is subject to liability to the extent it would be liable if it were a private entity.

Conclusion

The decision whether to use silent video surveillance technology in certain environments/situations, or decisions regarding the extent of its use raise serious constitutional law, privacy and negligence law questions. The wide array of technology that is available certainly encourages the use of video security systems. However, the design and implementation of any such system – and its periodic enhancement – must recognize legal parameters that both limit and expand liability. Evaluation of policy, staffing, training, and budget issues are essential, and should involve all administrators having responsibility for security. The absence of much case law does deprive us of the administrative efficiency we seek in the law, but the case law that does exist gives guidance that is quite rich in common sense and applicable constitutional and tort law principles that can be analogized with minimal intellectual uncertainty. The challenge is whether the law enforcement unit, government agency or employer is willing to devote serious planning, and budget (for training, staffing, maintenance, enhancement, *etc.*) to this aspect of policing/security, and whether a shared commitment to the appropriate use of technology may be obtained.

Sample Case Study

The Security Industry Association has recently completed the fourth and final volume of its CCTV for Public Safety series. This series, among other items, has tracked the successes and failures of CCTV use in public safety applications throughout the United States. The following is a sample case study of a public schools systems use of CCTV technology.

NEW YORK CITY, NEW YORK

Background

Since July of 1997, five of the most crime ridden public housing facilities have been participating in a pilot program exploring the value of closed circuit television monitoring. Around the time this project launched, New York City began a massive undertaking to rid crime throughout the city. It was hoped that the cameras would deter and reduce crime in these public housing buildings as well as enhance the overall security for residents and housing staff.

Public Safety Surveillance Application

The New York City Police Department is responsible for monitoring the equipment 24 hours a day, 7 days a week. Over 100 cameras are deployed throughout the five buildings. Cameras are located at building entrances, exits, rooftops, elevator cabs, lobbies, common areas, parking lots and perimeter locations. Three sergeants and nine police officers are assigned to each monitoring station, however the video footage is also recorded for later review. Each building has a separate monitoring station.

As the pilot program continues, the Housing Authority is looking into ways to enhance the capabilities of the current system. Soon, according to Millie Molina of the New York City Housing Authority, "video signals will be transmitted via telephone lines using a digital system. The distribution of high-density information over local area networks will enhance the central station operator's surveillance capabilities. In addition, a demonstration remote video monitoring system using NYCHA's computer network as a method to carry CCTV video signals is currently being prepared for possible installation." They are also hoping to develop a system to allow for monitoring of all buildings from one central station.

Cost

The CCTV system is funded by the Comp. Grant Program – Capital Funds Program and the modernization programs offered by the Department of Housing and Urban Development. The police who monitor the system are financed through the NY Police Department. Maintenance is covered under NYCHA operating funds. Since the

program's inception the city has average about $30,000 per year to maintain the program. Most of the costs going towards overhauling the time lapse video recorder after 10,000 hours of use.

Results

During 1999, New York City reported a 13.3% decrease in crime. Impressively, four of the five public housing units, which used CCTV, reported even a greater decrease. Since its inception, over 300 arrests can be attributed to CCTV through either direct observation or identification of perpetrators after reviewing videotape.

Program-At-A-Glance

Site:	**NYC Public Housing**
Date Installed:	**July 1997**
Previously Reported:	**1998**
Funding Source:	**HUD Grants**
Hours of Surveillance:	**24 Hours**
Implementation:	**NYC Housing Authority & Police Department**
Main Problem Addressed:	**Drug Dealing & General; Crime**
Results:	**300 arrests attributed to CCTV**

Resources

The *2001 CCTV For Public Safety and Community Policing* Report is in the final stage of completion as it moves from printed version to Internet and CD-ROM version. It became evident as the report neared 800 pages that it was in the best interests of the project that the report be complied in such a way as to facilitate Internet and CD-ROM dissemination.

The 2001 report is a compendium of all aspects of the project as specifically relates to:

- The CCTV Guideline;
- Review of the CCTV Summit;
- Review of Sample US cities currently using CCTV technology in public safety and community policing applications;
- Legal Issues related to Video Surveillance;
- Legal Briefs on CCTV Use;
- Samples of Federal and State CCTV Legislation categorized by Security Application;
- Review of Federal Grant Programs that Fund CCTV Programs; and
- Related Supplemental Program documentation listed in a detailed Appendix.

It is envisioned that this compendium will be accessible on the SIA and IACP websites as well as be available on CD-ROM in a PDF format and/or searchable format, free of charge or with a nominal postage/handling fee. By placing this compendium on the Internet SIA/IACP will be able to regularly update the project and provide up-to-date information on a more regular basis and thus extend the life and effectiveness of the project.

The next iteration of this program will be to establish the statistical effectiveness of the CCTV technology as a public safety and community-policing tool. Discussions are underway with representatives from Stetson University and The University of Tampa to conduct this research. SIA is willing to fund the project and the compilation and dissemination of the data for the PSLC. It is expected that this statistical part of the program will be completed in late 2001 mid 2002.

For more information on responsible CCTV use in public safety applications, contact Richard Chace at chace@siaonline.org or visit www.securitygateway.com.

Appendix B

Selected CCTV Related Statistics

SIA
SECURITY INDUSTRY ASSOCIATION

CCTV Crime Statistics

Studies done in Stevenage and North Hertfordshire, England

- Overall 30% reduction in crime for three years immediately after implementations of CCTV (page 1).
- CCTV responsible for the discovery of 7,000 incidents leading to nearly 1,000 arrests in three years (page 1).

Study conducted in Belfast, Northern Ireland

- Thefts from vehicles decreased 31.8% compared to 23.3% in non CCTV areas (page 2).
- 50% increase in shoplifting cases otherwise gone undetected (page 2).

Study conducted in Lincoln, England

- Crimes to vehicles in a parking garage decreased by 85% when CCTV cameras were installed (page 14).
- Trends show that after CCTV is installed there is an immediate increase in the report of crime, but in half of the cases crime decreases in these areas after a year (pages 20-22).
- CCTV is effective in identifying and deterring vandalism, robbery, violence, and public disorder incidents, also increasing police response times (page 22).
- High Street - Before CCTV cameras were introduced the average number of crimes per month was 44.5. One year after installation of CCTV cameras this number was reduced to 18 per month (page 18).
- High Street - Represented 9% of the district crime figures, and after one year of CCTV in use-only 4.4% (page 18).

Evaluating and Monitoring CCTV-Peter Fry, National CCTV users group

- In a survey conducted prior to the installation of a CCTV system, 30% of respondents felt safe in the areas of highest crime. When asked the same questions one year after the CCTV installation, nearly 80% felt safe in those same areas (page 27).
- The same survey revealed that prior to CCTV installation, over 70% of respondents had witnessed acts of crime. The second survey after the CCTV system installation showed that nearly 60% witnessed no crime at all (page 27).
- Crime reduction on Industrial Estates with CCTV increased dramatically in the first two years, reaching approximately 40%, and a then overall decrease in crime in the next eight years (page 29).

Security
Industry
Association
635 Slaters Lane, Suite 110
Alexandria, Virginia
22314-1177
Telephone: 703.683.2075
Facsimile: 703.683.2469
www.siaonline.org
www.SecurityGateway.com

PRIVACY VS. SECURITY: ELECTRONIC SURVEILLANCE IN THE NATION'S CAPITAL

HEARING

BEFORE THE

SUBCOMMITTEE ON THE DISTRICT OF COLUMBIA

OF THE

COMMITTEE ON GOVERNMENT REFORM

HOUSE OF REPRESENTATIVES

ONE HUNDRED SEVENTH CONGRESS

SECOND SESSION

MARCH 22, 2002

Serial No. 107–166

Printed for the use of the Committee on Government Reform

Available via the World Wide Web: http://www.gpo.gov/congress/house
http://www.house.gov/reform

U.S. GOVERNMENT PRINTING OFFICE

85–122 PDF WASHINGTON : 2003

For sale by the Superintendent of Documents, U.S. Government Printing Office
Internet: bookstore.gpo.gov Phone: toll free (866) 512–1800; DC area (202) 512–1800
Fax: (202) 512–2250 Mail: Stop SSOP, Washington, DC 20402–0001

COMMITTEE ON GOVERNMENT REFORM

DAN BURTON, Indiana, *Chairman*

BENJAMIN A. GILMAN, New York
CONSTANCE A. MORELLA, Maryland
CHRISTOPHER SHAYS, Connecticut
ILEANA ROS-LEHTINEN, Florida
JOHN M. McHUGH, New York
STEPHEN HORN, California
JOHN L. MICA, Florida
THOMAS M. DAVIS, Virginia
MARK E. SOUDER, Indiana
STEVEN C. LATOURETTE, Ohio
BOB BARR, Georgia
DAN MILLER, Florida
DOUG OSE, California
RON LEWIS, Kentucky
JO ANN DAVIS, Virginia
TODD RUSSELL PLATTS, Pennsylvania
DAVE WELDON, Florida
CHRIS CANNON, Utah
ADAM H. PUTNAM, Florida
C.L. "BUTCH" OTTER, Idaho
EDWARD L. SCHROCK, Virginia
JOHN J. DUNCAN, JR., Tennessee
——— ———

HENRY A. WAXMAN, California
TOM LANTOS, California
MAJOR R. OWENS, New York
EDOLPHUS TOWNS, New York
PAUL E. KANJORSKI, Pennsylvania
PATSY T. MINK, Hawaii
CAROLYN B. MALONEY, New York
ELEANOR HOLMES NORTON, Washington, DC
ELIJAH E. CUMMINGS, Maryland
DENNIS J. KUCINICH, Ohio
ROD R. BLAGOJEVICH, Illinois
DANNY K. DAVIS, Illinois
JOHN F. TIERNEY, Massachusetts
JIM TURNER, Texas
THOMAS H. ALLEN, Maine
JANICE D. SCHAKOWSKY, Illinois
WM. LACY CLAY, Missouri
DIANE E. WATSON, California
STEPHEN F. LYNCH, Massachusetts
———
BERNARD SANDERS, Vermont (Independent)

KEVIN BINGER, *Staff Director*
DANIEL R. MOLL, *Deputy Staff Director*
JAMES C. WILSON, *Chief Counsel*
ROBERT A. BRIGGS, *Chief Clerk*
PHIL SCHILIRO, *Minority Staff Director*

SUBCOMMITTEE ON THE DISTRICT OF COLUMBIA

CONSTANCE A. MORELLA, Maryland, *Chairman*

TODD RUSSELL PLATTS, Pennsylvania
THOMAS M. DAVIS, Virginia,
——— ———

ELEANOR HOLMES NORTON, Washington, DC
DIANE E. WATSON, California
STEPHEN F. LYNCH, Massachusetts

EX OFFICIO

DAN BURTON, Indiana

HENRY A. WAXMAN, California

RUSSELL SMITH, *Staff Director*
HOWIE DENIS, *Professional Staff Member*
MATTHEW BATT, *Clerk*
JON BOUKER, *Minority Counsel*